Working on a Farm

Alan A Mister

Aberdeen Angus bull: famous for high quality beef

Batsford Academic and Educational Limited
London

First published 1982

Typeset by Progress Filmsetting,
79 Leonard Street, London EC2A 4QS
and printed in Great Britain by
Butler & Tanner Ltd
Frome, Somerset
for the publishers
Batsford Academic and Educational Limited
an imprint of B T Batsford Limited
4 Fitzhardinge Street, London W1H 0AH

British Library Cataloguing in Publication Data

Mister, Alan
Working on a farm—(Careers series)
1. Agriculture as a profession—Juvenile literature
I. Title II. Series
630'.2'03 S494.5A4
ISBN 0-7134-3962-9

Contents

Acknowledgment

I would like to thank all those who have made it possible for me to write the book. Firstly, to Peggy, my wife, who has spurred me on when my spirits have sunk low. Certainly next, my thanks and gratitude to the Agricultural Training Board, and in particular to those of its staff who love the land and seek to serve the interests of those whose labours ensure it remains fruitful; the National Union of Agricultural and Allied Workers in whose service I was able to meet so many skilled and dedicated members of the work-force; the Soil Association, who for so many years have held high the torch of organic farming and, especially, those of its members who show in their daily toil that natural methods work. To Charles Pinney of Cotswold Park Farm, whose efforts to assert the place of the Heavy Horse in our farming scene are so wise and far-sighted when oil is not only costly but its scarcity threatening the peace of the world; to that numberless company of friends who by talking to me, and with me, have filled the gaps in my knowledge which would otherwise have been such a hindrance in writing this book. When describing 'Our Farming Scene', these gaps were all too apparent to me and it was with some relief that I recalled Leslie Thomas's excellent booklet, *Farming in Britain,* published by the Association of Agriculture. In dealing with those regions of Britain with which I am not too familiar, I found rereading this enabled me to fill them adequately—thank you Leslie Thomas and the Association of Agriculture.

In two essential matters, my glaring ignorance sent me flying to expertise to fill the vacuum. In bee-keeping my rescuers were Read Admiral Tom. W. Best of Hincklowe Fruit Farm, Melplash, Dorset, Ben Kardas of Dorchester and Arthur Worth of Old Stafford, both in Dorset. Any intending bee-keeper would learn much by sitting at the feet of these three gentlemen, to whom I am indebted.

When it came to advising on agricultural college training the glare became blinding. Groping my way to Dorset College of Agriculture in Hardy's parish of Stinsford, the gaps in my knowledge were filled with great patience and kindness by the principal, Mr H F Fieldsend, B Sc (Agric) and the organiser of part-time classes, Jim Wilson B Sc (Agric). Not the least of helpers in the rich agricultural county of Dorset, was Alan M W Roberts BA, who, as county secretary of the National Farmers' Union, keeps the sights of the farming community on the high plateau of vision and yet found time to comment on the reference made to his organisation in this book.

Dorchester 1982 AM

Captions to photographs

National proficiency test, milking and dairy hygiene

Caring for the feet of cattle

Power take-offs

Attaching a three point linkage for ploughing

Horticultural demonstration garden with teaching plots in the Elizabethan walled garden at Kingston Maurward Manor

Dagging is an important task with sheep (using hand shears)

Instruction in amenity horticulture

Instruction in electric arc welding

The above photographs reproduced by kind permission of the Dorset College of Agriculture, Kingston Maurward, Dorchester.

Sorting stones from potatoes which have been lifted by a mechanical potato harvester

Cabbage cutting

Foot trimming of sheep

Clearing slurry

The above photographs are reproduced by kind permission of the Agricultural Training Board.

Drawings are reproduced by courtesy of the Association of Agriculture.

Cover photograph reproduced by courtesy of Massey Ferguson, Coventry.

Preface

The land of these Islands has been described as a heritage and the element from which our roots as a nation derive their sustenance. It is much more: it is our only renewable resource, provided, of course, that we care for it in accordance with the basic laws of husbandry—a word you will come across frequently if you decide to work on the land.

A 100 years ago the majority of the population was either engaged in farming or those rural crafts that served the industry. Even now many of those living in our towns and cities would admit to some link with the living soil and that part of the animal kingdom we call farm stock. With the birth of the Industrial Revolution the material standard of life of ordinary people gradually rose. Men and women flocked to the towns where the streets were said to be paved with gold. As our manufactured goods flowed to the four corners of the globe, so more and more imported food poured in through our ports, and it was not until the Great War of 1914 to 1918 that agriculture regained some of its former glory. But only for a short time. By the 1920s the depressed conditions returned.

With the outbreak of the Second World War farms had once more to be restored to their former well-being. The nation as a whole realised what an important stake it had in food production. This was reflected in new Legislation aimed at guaranteed prices for the producer and a stable economy from which land owners, farmers and those employed on the land could derive a real sense of security.

This book will give you an insight into the many facets of work on a farm, and, I hope encourage you to make farming your career.

An introduction to the world of food growing

In agriculture, unlike industry and commerce, there is no overall pattern of career structure, though if you mean to obtain a foothold you will find in the pages that follow a clear pattern of opportunity. Once you have made the decision to get a job in food growing your own position will emerge. From the word 'go' you will find in your work opportunity, comradeship, satisfaction, and above all else, happiness.

There are, however, more jobs in farming for a young man than for a young woman, but girls have made a very special contribution to the industry as workers and, as employers. Any girl with the will to succeed is likely to surmount any obstacles that may stand in the way of her career. Horticulture, in particular, is dependent on the special qualities of the female labour employed—who form the majority of the total work-force.

Farming is 'a way of life' and certainly, within the restraints of job discipline, there is an opportunity to do one's own thing. The unpredictability of so much that happens in one working day on a farm will guarantee an interesting and varied job.

Of the school-leavers entering full-time work in farming each year less than a quarter go into the Apprentice Scheme. As a life-long believer in training I come down unhestitatingly on the side of those in the industry who want this Scheme to work and, more importantly, see in it the most sensible way to ensure that farming in this country is served by a work-force well able to ensure its position as one of the most efficient and best in the world.

As more than three-quarters of recruits to the industry enter it from outside the Apprenticeship Scheme, chapter 4 may be of particular interest. Here I have tried to cover all possibilities, with particular emphasis on opportunities outside the mainstream of commercial agriculture, because I believe changing forces affecting farming, both here and abroad, will reassert the role of the smaller type of holding and recognise the need for more job opportunities to be created so that rural life, as we know it, can survive.

1 Our farming scene

There is indeed so much variety in the soils of Great Britain that it would be easier to describe the entire landscape of Denmark than the variations to be found in one English parish. This may be an exaggeration, but the kernel of truth it contains will help you to grasp the three basic factors in farming: soil, climate and location, and, from a brief trip around the regions it will be easier to understand how soil and weather affect the agricultural pattern. Have an atlas handy so that you may follow our journey with more enjoyment.

The South East

The South East is bounded on the north by the low-lying estuary of the river Thames while south and west, from the North Foreland to Southampton Water, the English Channel is the boundary. The New Forest and the edge of Salisbury Plain mark the fertile, arable western boundary with the chalk uplands of the Chiltern Hills forming a northern rim. Everywhere in this area the spring comes early. Much of the region has well-wooded uplands. The Weald, Ashdown Forest and the ancient Silva Anderida, or what is left of that forest well-known to the Romans, are important recreational areas serving an enormous urban population. Between the North Downs and the Thames is the rich fruit and market garden district which together we call the Garden of England. As well as bearing this enviable title, Kent has three times more land under fruit than any other English County. Increasingly, the older, high spreading fruit trees are giving way to intensively-planted, smaller bush-like trees. Also in Kent, on the deeper soils, as well as in East Sussex and Hampshire, two thirds of this nation's hops are grown.

West Sussex, being blessed with a high sunshine and reasonably pure air, supports a flourishing glasshouse industry where tomatoes, cucumbers, salad crops and a great variety of beautiful flowers can be grown more cheaply than in less-favoured districts where sunshine is elusive and the levels of air pollution much higher.

This is the region of much *mixed farming,* the term used when arable cultivation and livestock farming, based on grassland, is carried on in the same holding. Not all the Weald is upland woodland, its clay soils are the scene for dairying and stock rearing. In south east Kent barley and wheat are extensively grown as they are in the chalk country of Hampshire and Berkshire. The sloping downland in these counties presents a grand picture in spring and summer from many high vantage points described so lovingly by H J Massingham in his *English Downland.* This gigantic patchwork of waving wheat and barley surrenders to the giant combine harvesters every autumn. Who knows, this satisfying job may come your way when you start on your farming career but not before you have proved your skill as a combine driver and learnt to deal with the minor troubles that can halt operations.

Like most of the southern England, in fact much of the country, dairying means the black and white British Friesian. This beast provides us with almost 80 per cent of our milk and almost two-thirds of our beef, more, if you include the carcasses of cross-breed cattle. Some of the grass in the south east is of poor quality, particularly in mid-Sussex, and it is here you are likely to see the sleek red Sussex cattle; surely these and the Devon and Aberdeen Angus beasts are the most handsome of all? Charolais, Simmental and other continental cattle have spread into the region and, for that matter, all over the country. You may find shaggy Highland cattle on the slopes of Ashdown Forest or at the bottom of Crowborough Warren. The Hereford, another noble animal is to be found here, crossed with the Friesian for beef production.

Sheep production is increasing in the south east, though is still not as extensive as in other areas. The South Downs have given their name to one of our supreme animals, which is in complete contrast to the Pevensey Levels or Romney Marsh, which have also named a world-famous sheep. The shearing of these animals' fleeces is now a feature of many of our Agricultural Shows.

Pig breeding flourishes in many parts of the region and until recently the Thames Valley, West of London, supported small holdings of this most intelligent of animals. The intensive production of pigs, or for that matter, any other animal, I find a degrading form of husbandry, but, alas, the economics of modern agriculture suggest it is justified to ensure a good return and enable farmers to remain solvent.

There is nothing like pigs to bring you down to earth! Still to be found in the region, is the handsome British Saddleback, the Large White, the Landrace, besides many of the smaller breeds now dwindling and being replaced by what are known as crosses. I cannot leave this, my first reference to the animal, without recalling its many qualities, all of them best seen when the creatures are kept in the open, able to 'rootle', a word used to describe the blissful activity of pigs as they seek to find their favourite foods, including acorns and beechnuts, but which would not produce the carcass now favoured by the British public.

Reference must be made to the vast industry of poultry rearing, common to the entire country as well as the south-east. Although free-range birds are still to be found, in fact their numbers are increasing, most job opportunities in this field are in controlled-environment buildings, as I have heard them described.

The South West

The true boundaries of the South West are debatable but I will always go for the area comprising the counties of Cornwall, Devon and Somerset. However, I must also include Dorset, Wiltshire and Gloucestershire as they cannot easily be included with another region.

Certainly, it must include the Scilly Isles, one of our greatest flower growing areas despite its small size. Early in the seasons magnificent blooms arrive as spring touches this corner of England first. Moving into Cornwall it enables early potatoes to be grown as well as the first strawberries and early broccoli. No description of the region can long ignore the thousands of hectares of grassland. Indeed, it is the foremost farming asset. There are many different grasses and these exist nationally making up those patches of green in our landscape, beloved by our countrymen and the envy of overseas visitors from less verdant lands. However, in themselves they comprise only a part of the green—many weeds, too, have a role to play as an animal tonic.

The South West, with its patchwork of small fields, particularly in Devon and Cornwall, is still a stronghold of small farmers, who survive because their stubborn love of the land often enables them to exist on a subsistence level of income. High rainfall in the upland farms rules out most crop production, hence the dependence on sheep, beef and dairy production. Devon, as well as being a great milk producing county, carries more sheep on its lush pastures than any other county except the Ridings of Yorkshire.

Holiday-makers think of this area as one of gentle loam soils and rich valleys but those working in it know how severe the winters can be and to live in a village in the hills in a bad winter, is an experience never to be forgotten, especially when blizzard conditions return, after a false thaw, to trap thousands of sheep.

Some of our best cheese is made from the rich milk of the West Country though few farmhouses make the Cheddar that brought fame to the district. However, there is evidence that farmhouse co-operatives, devoted to the manufacture of a high quality product, are on the increase again and it is to be hoped that this trend will develop and provide more rural work opportunities.

No region has more varied patterns of farming than the South West and none is more pleasing to the eye. One locality of special interest is that massive saucer of low-lying land—the Somerset Levels—lying between the Quantock and Blackdown Hills to the south, and the lime stone of the Mendips to the north, which is a major source of the county's milk production. Much lies below sea level and only careful land-drainage prevents it reverting to the swamp and sedge of former centuries.

Here, also, is the home of the willow osier, from which the coppiced reeds are cut and the old craft of basket-making carried on. Even to this day, some of the fine cider, or 'scrumpy' is still made here, while under the Mendips, strawberries and vegetables are grown.

Leaving Somerset, we travel into the chalklands of Dorset and Wiltshire, the former becoming again one of our foremost sheep counties. In both counties large farms are found and towards the south-east and adjoining the rash of urban development around Bournemouth, another horticultural locality based on the light sandy soil. While still producing run-of-the-mill crops, the Vale of Pewsey, Wiltshire, has thousands of hectares under grass—seed for re-seeding—for establishing new pastures or leys.

The large farms are growing more and more oil seed rape. When you see enormous fields a mass of yellow in late spring and early summer you will be looking at this crop in an important stage on the road to maturity. It is grown under contract for the production of animal protein and becoming a common feature of lowland arable farming.

Pig production for supplying processing factories is carried on from intensive holdings in central Wiltshire and the end products—chops, sausages, hams and pies—are deservedly famous, being marketed nationally under equally famous brand names. In the south and north of the county are large market gardens and fruit growing areas, committed to the survival of the peerless English dessert apple, now threatened by competition from across the Channel.

Gloucester, the home of a great wool industry in the Middle Ages, has three main farming systems. The first is found on the cold Cotswold country, based on large-scale cereal growing and grass leys. The cereal acreage continues to increase year by year and as the other vital factor, milk production, is now tending to decline, this increase is likely to become more marked and fat lamb production should show an increased buoyancy. The second system is one of producing as much milk as possible from as many cows as the land will support on the small farms of the Severn Vale. Each year more and more small farmers give up the struggle and surviving holdings continue to grow in size. Lastly, west of the Severn is the Forest of Dean. Here are large areas of unenclosed sheep grazings, numerous small dairy and stock rearing farms while men employed in local industries maintain part-time holdings.

As I have gone outside the West Country border, perhaps another lapse would be permissible to ensure a reference to the Channel Isles and the Isle of Wight. With much of the South West, these islands are providing more and more members of the farming community with a most reliable 'crop'—tourism. More will be written of this new role for farmers but as far as this chapter is concerned all that need be said is that people will come to holiday on farms because they are attracted by what goes on. Hence, the branch of tourism building up within the farming scene will increase in importance as individual farms, catering for visitors, provide a background of really first-class husbandry, which means the need for skilled labour.

The Isle of Wight is mainly dairy in the north while in the rolling south cropping and livestock rearing are carried on. In the Channel Islands one first thinks of early potatoes, tomatoes, the hardy Jersey cow, with its high butter-fat content milk, the heftier Guernsey breed with equally rich milk, and the small wine producing industry.

Our tour through the South West must include a reference to the job opportunities there. Although over the last 30 years the farm labour force has declined from nearly 60,000 to about 27,000 the story is not all black. More and more job opportunities have developed in rural industry and forestry and it is to be hoped that ways and means will be found to stabilise a reasonably sized work force throughout the region.

The Midlands

The Midlands form two areas for the purpose of our journey through the farmlands of Britain. The West, stretching from the Marches of the Welsh border to Birmingham and the East Midlands, from the one-time Lindsey Division of Lincolnshire in the north-east to the top of Buckinghamshire in the south. The remainder of the county of Buckinghamshire, together with much of *Oxfordshire,* belongs to south country farming. Up in the north, which most people think of as The Midlands, lies the high and wet Peak District of Derbyshire.

Milk, sheep and beef sums up West Midlands farming. Over 200,000 hectares of cultivated land is cropped for grass; the temporary, or short term leys, which will always be the mark of good husbandry. Here, too, the grass-land management carried on ensures good sheep and beef cattle production of a very high order.

A sight to bring joy to any would-be farmworker is the thousands of sheep changing hands at the great autumn sales in the Marches. Around the market town of Craven Arms, the great hills of Long Mynd, Caradoc, Wenlock and Clee provide the home pastures from which flocks come in large numbers. Among them the black-nosed Kerry Hill, the Shropshire and Radnor Breeds and, dominating them all, the Clun Forest, a sheep deservedly popular for its thick and yet delicately-textured fleece. The great contribution of Herefordshire to our farming scene is the Hereford, the most popular breed of all cattle for the production of fine beef. It does well on almost any pasture, and on terrain in some parts of the world that hardly justifies the description.

Together with Worcestershire, Herefordshire produces over a quarter of the nation's hops and, apart from fruit growing, here, too, can be found large production of that aristocrat of vegetables—asparagus. In Shropshire we find to the north, and also in adjacent Cheshire, another area famous for cheese-making, but today only on a small number of big dairy farms.

Milk is the main product in Cheshire and the Friesian is the most popular dairy cow here, as it is all over the country nowadays. Nevertheless, most of our world famous breeds will continue to survive against its competition for many more years. Milk is the main farming theme in neighbouring Staffordshire. It possesses a varied pattern of farming, including live-stock rearing, considerable arable farming and some horticulture in the south. Here, too, is where the region's main producers of barley, potatoes and sugar beet can be found; the British Sugar Corporation maintain processing factories, which are a source for rural job opportunities.

Warwickshire is now the permanent home of the country's premier farming show, the 'Royal'. On the showground at Stoneleigh more and more activity related to modern agriculture is blossoming. One such development is the Centre of the

Agricultural Training Board, where specialist courses in farm management are offered within the industry. Here, hopefully, many farmers and workers are learning that difficult skill, the art of togetherness and mutual aid for all the staff of a farm.

Returning from the Royal last year I was able to grasp the delicate balance in the land use in this county. Industry continues to make great demands on farm land and those fields that remain are required to somehow produce the same amount of food, or even more. How long this tight-rope can be walked without catastrophe is something we should all consider. I was also able to view the general farming scene and note some fine examples of an old rural craft, hedge-laying. Many holdings of over 4000 hectares are to be found on the heavy wheat land, extensive sheep farming, as well as small family farms.

The eastern part of The Midlands has a lower annual rainfall and bleaker winters, which can be very wet. Only ten years ago heavy land was being left uncultivated in a run of wet winters. Despite the problems, it is an area devoted to cereals, potatoes and sugar beet, as well as extensive grassland, including the world famous fattening pastures in South Leicester, and around Market Harborough in Northamptonshire.Here too are found the great areas of intensive field-scale production of horticultural crops. In Bedfordshire the unmistakable smell of Brussels sprouts, denotes generations of vegetable growing. In much of Lincolnshire and, to a lesser extent in Nottinghamshire, what can be termed the garden vegetables are grown on contract to the strict conditions of the major food processors and distributors.

Wales

We now pass into Wales, already glimpsed from the high country of the Marches, as well as across the wide reaches of the Severn. To travel anywhere there, is to take a ride on a switchback. I first entered the county while seeking the source of the River Wye and in my visits since then I have formed an impression of great hills, often menacing, as the rain and snow clouds sweep in from the West.

Fertile lowlands are to be found in the South of Pembroke, now called Dyfed, a narrow coastal belt at Colwyn Bay and, of course, in the Gower peninsula. In Gwent (Monmouthshire) there is more lowland farming. As well as the rain (up to 2.54m a year on the mountains) the farmers have to cope with extensive bogs. Sheep rearing is the order of the day and what a varied pattern we find.

Border Leicesters are crossed with the mountain ewes, Downland rams like the Suffolk, Southdown, Dorset and Hampshire Down also fit into this pattern.

The varieties of sheep to be found in Wales poses problems which farmers have to face, and much research is carried on in connection with what are termed genetic factors, as well as into the nutritional needs. The Welsh Plant Breeding Station, Aberystwyth and the Pwllperian Experimental Farm nearby are involved in this research and graze sheep right up to the mountain tops; the latter farm also has Welsh Black cattle grazing on high, improved pastures described as 'looking like golf links in a tundra landscape'.

In Wales the role of forestry within the farming framework is both obvious and vital. In chapter 13 forestry is explained as are the job prospects it provides. What has been done in Wales by the Forestry Commission, and by private forest owners, has helped the hill farmers. The steep slopes have trees clinging to them, often with the bare minimum of soil. The drainage undertaken before planting and the making of access roads all play their part in pasture improvement that would otherwise be uneconomic.

The Welsh Black cattle, the other valuable inhabitants of the mountain country, are famed for their mother role. By suckling calves in trying conditions they provide an important source of young stock that can be fattened in the lowlands. This is also done with the Hereford and Hereford-crossed beasts in Brecon, Radnor and Powys, the partly reared calves being taken away to the Midlands and other counties of England.

Back in Dyfed, Anglesey and the Gower Peninsula early potatoes are grown to augment the supplies from English counties and, in fact, to compete on the market with them. Kale, a crop not mentioned before in our travels, together with rape and barley, are grown for livestock; the cattle and sheep grazing them before the spring grass appears. To the Principality's credit it should be recorded that 10 per cent of the combined output of the two countries is produced there, including almost one and a half million gallons of milk.

The North of England

A region of unique physical features; the climate, too, plays a major part. Doubtless, you have learned that the Pennine range of hills, aptly called the Backbone of England, form a natural watershed—a boundary between two distinct weather patterns—

the west being the wet side, with drying conditions prevailing to the east. In these times of climatic upheavals and topsy-turvy weather it is possible that the division is less clear than in former years. The Cheviot Hills which form the natural border with Scotland, serve as an askew top of the tee to the Pennines while in the Lakeland mountains we have, possibly, the most intriguing part of the region.

This area, because of its beauty, is an ever-popular holiday area. Each year more and more visitors squeeze into it, and the remaining parts of Cumbria. As in the West Country, the holiday trade is a vital source of income to the farming community, at the same time, this important cash crop poses a threat to the important agricultural pattern. Swaledale sheep and other hardy breeds, as well as cattle, graze the all-prevailing grass, where a severe climate, marked by heavy rainfall, gives the farmer sufficient challenge without the added anxiety of gates being left open and stock being harassed.

Moving on to Northumberland and Durham, where the main agricultural enterprise is still hill farming, there are few beasts to these harsh hectares. Autumn sales play an important part in disposing of any superfluous stock that might threaten a holding with over-grazing and major winter feeding problems. Here, as in the Lakes, forestry is a major enterprise with the largest man-made forest in Europe sited on the Northumberland-Scottish border.

It has been described as covering the equivalent of more than a 1000 average-sized commercial farms.

Lancashire brings us back to large-scale dairy farming and, in the south west of the county, to intensive market gardening. This rich land is best seen when flying from the south to Blackpool. The flight path taken invariably crosses the serried fields of vegetables used in the adjacent urban areas. Horticulture is also very important across the Pennines, in the Vale of Pickering, where it is combined with grain growing on the drier, well-drained deep soils of Yorkshire.

To attempt a fair appraisal of the farming scene calls for the loving knowledge of a native. Winifred Holtby did it a generation ago in *South Riding* and other novels—read them if you want to get the feeling of a landscape that changes only with the passage of time.

This North Country survey should include a reference to the rearing and fattening farms north of Newcastle and the very good mixed farming to be found between the rivers Tyne and Tees.

The Eastern Region

The Eastern Region as it is described by the Ministry of Agriculture, is the driest in the country, and in places returns rainfall figures that should classify them as drought areas. Extremes of temperature and the level of bright sunshine are above the national average. Irrigation is an important fact of life most years and if you get involved on the land here you will soon come to grips with the chore of getting water to crops that would otherwise wilt, sicken and die for lack of moisture.

Always a windy region, especially in those parts where hedgerows are non-existent, the high returns on corn and crop growing virtually rule out any large-scale tree planting. This has brought into prominence the tendency of fine, dry, dusty land to 'blow' and encourages the view that unless more consideration is given to varying the husbandry pattern something resembling the Dustbowls of the Mid-western States of American could occur here.

The sea has an important moderating effect on large areas around the coast, and the many sea inlets and estuaries, particularly on the Essex coast, have a marked influence on the frequency of frost—a vital weather factor in fruit-growing and horticulture. Even a brief description of the climate of the region would be incomplete without reference to the easterly winds that blow persistently well into spring and the bitter north-easterly can seem as frequent as the prevailing westerlies.

The mention of geology should never deter a would-be landworker. Certainly the soil structure is affected by the subsoil beneath and for many centuries it was an important skill to be able to run cultivated soil through the fingers and hands in an effort to size up its food-growing ability as well as its friability. These days soil analysis can be undertaken both by the farmer and the Ministry of Agriculture Advisory service so inspired guess work is out.

The region's soil structure is bound up with the geological formation. The cropping of these soils is a source of study at the *Arthur Rickwood Experimental Husbandry Farm* in the *Cambridgeshire Fens* where many challenges are being wrestled with. In particular, the importance of the liming programme on sandy soils, to ensure high yields in barley and sugar beet, the effect of drought on the light loams, the low stability in silts, so valuable for growing root crops, the drainage factor in boulder clays and, finally, the rapid loss of organic matter that can occur in

peats.

More than anywhere else in the British Isles, East Anglia is dependent on efficient land drainage. A quarter of the Region is low-lying and depends entirely on man-made land drainage for viability. So effective has this work been since the Dutch engineer, Vermuyden, started tackling it in the seventeenth century, that more and more land has come into arable production whereas it had previously been suitable only for summer grazing. Most of the back-ache has now been taken out of the work by using tackle attached to wheel and crawler tractors but patching up and emergency jobs still call for skill and brawn to keep the water flowing.

Along the Thames estuary, the region's southern boundary, and the North Sea coastline, large areas of both arable and pasture land lie below sea level. This calls for an effective sea defence system, protection being ensured by the provision of sluices which open only at low tide. In 1953 a tidal surge swept all before it, flooding large areas, destroying buildings and causing tragedy and the salinating of farmland. In Suffolk and Norfolk there is the additional problem of erosion by the action of the waves, while in The Wash, where Norfolk joins the rich Spalding area of South Lincolnshire, the sea is in retreat and good farmland is being won by enclosing the foreshore within earthen embankments.

This being a region of gently undulating land, large scale, mechanised farming comes into its own. Here, over two million hectares produces arable crops, being almost a quarter of the total arable land in England and Wales. Understandably, this large-scale crop production has led to the establishment of back up services—agricultural machinery manufacturing, fertiliser and spray production and widespread marketing facilities.

All play their part in the production of a third of the nation's wheat and potatoes, two-thirds of its sugar beet, half the feed bean and vining pea crops and 80 per cent of dried peas. All this farm activity has been affected by an acute manpower shortage. To the lover of pastures and stock-raising, the low rainfall makes grass production unreliable and permanent pasture is rarely found. However, as pasture has been increasingly buried by the plough and the land used for cereals, etc, so have grass weeds become a problem on many farms. The ever-increasing cropping of cereals has also brought about more infection from disease—mildew, yellow rust and eyespot—both these enemies of intensive farming are combated by the increasing use of herbicides and fungicides.

High prices caused by the world demand for vegetable oils,

coupled with an assurance of price stability to the grower, has resulted in a large increase in oil seed rape growing and despite the small grassland areas, the region produces over a third of the nations' grass and clover seed as part of the arable rotation.

Horticulture takes its place alongside farming in terms of crop production, in fact the two are interwoven. In the Eastern Region about half the nation's vegetables are grown—Bedfordshire for Brussels sprouts, though more and more are being grown in Norfolk: Essex is a major sweet corn producer, Cambridgeshire is the country's main celery producer while in Hertfordshire watercress growing is very important. Norfolk produces great quantities of parsnip, asparagus and mint for processing. Suffolk, too, grows much asparagus.

Glasshouse crops and top fruit complete the picture and despite the fame of both Kent and Worcestershire for growing apples, about 30 per cent of the national dessert apple crop comes from the region, mostly grown around the Colchester area, though small areas can be found in Norfolk and Cambridge. The Wisbech area grows our famous Bramley cooking apple while Comice and Conference pears are cropped mainly in Essex and Suffolk though Wisbech, too, makes an important contribution.

The region has 15 per cent of the national bulb growing area and the Spalding Flower Festival provides a unique, annual occasion for showing the beauty of the end product—narcissus, tulips and other flowers being used, seemingly by the million, to create the floats that make up the parade. More and more bulbs are going to the dry bulb trade and interest in the export trade is on the increase, growers striving to reach the high standards set by the importing countries.

The large scale production of crops in the region is 'intensive'. Although the year by year balance sheets of growers would suggest such husbandry is efficient, this is not the whole story. Crop diseases and plant pests are an additional hazard to add to the problem of weeds already mentioned when cereals are grown in old pasture land. An army of scientists now deal with this problem nationally and, nowhere are they more busy than in the Eastern Region.

To date, most of the research has gone into the development of more complex and powerful sprays but in the future more effort will be made to develop strains of predators to live off the flies, caterpillars, moths and spiders, etc, that plague plants and fruit. The natural balance of nature ensures that this happens to some extent, a well known example being the ladybird, to be seen scuttling around plants from late spring onwards and nowhere happier than on rose bushes, eating up large quantities of aphids and greenfly. Certainly, new techniques will have to be found as more and more of our pests are developing a resistance to sprays. In the Lea Valley, biological control methods, as they are called, are being adopted.

The region is not entirely given up to the production of cereals, root crops, vegetables and fruit. The production of livestock is vital and considerable. For example, in recent years areas, where sheep rearing had gone out of existence, are now witnessing the appearance of small but increasing flocks. About 20 per cent of the United Kingdom pig herd is kept in the region.

Most of the pigs are Large Whites, Landrace and Welsh but 'hybrids' or cross-bred herds are on the increase with artificial insemination taking the place of the natural function of procreation.

After declining in numbers for many years, the last ten years has seen a rapid growth in the number of beef cattle in the region. The main reason why the trend was reversed was expansion in suckler herds. These have been developed on farms with some permanent pasture and with arable crop residues such as beet tops, sprout stalks, carrots, etc, for winter feeding. Herefords crossed with Friesians are in the forefront in the newly established herds found in most of the counties right up to the London boundary. Certainly, an observant traveller would find such herds, as well as high quality milking herds, within ten miles of Charing Cross.

Sugar beet tops should not be confused with the fodder beet being grown in small acreages for herds, principally in Norfolk, while maize silage has proved very palatable on several farms with beef herds. This brief survey of feeds would be incomplete without a mention of barley which was first introduced in the early sixties and has been on the increase ever since.

Less than 5 per cent of our national dairy herd is to be found here and such herds as there are seem to be getting bigger all the time. It has been my experience that very large herds, of between 100 and 200 beasts, pose special problems and tensions. These are frequently reflected in staff changes. The average herd size, however, is 66 cows as against a national average of 46, so herdsmen are not on the move all the time. Most milking is done in up-to-date 'parlours', as they are called, and a high standard of cleanliness with both equipment and buildings has to be maintained. *Remember always cows have to be milked 14 times every week of the year.*

Sheep production, has only been possible in an area of great arable specialisation, because temporary leys have been sown, particularly on light land farms. Pedigree sheep breeding is, however, almost unknown apart from a small number of Suffolk breed flocks. Lambing tends to be early, and top prices obtained as a consequence, but great ingenuity has to be shown by flock owners in fitting into the feeding rotation, forage crops that will only occupy the ground for part of the year. There is always the possibility that grazing on the short-term leys will be curtailed by the all-too-frequent summer droughts.

The poultry industry accounts for a very substantial part of the region's agricultural output. Broilers, turkeys and ducks being prominent in the table section. Egg-laying stock is kept on a large scale and flocks of poultry tend to be large, 54 per cent of egg-laying flocks, are more than twenty thousand in number. Over two-thirds of the broiler fowl population is in flocks over 100,000. Norfolk has the largest share with Suffolk and Essex sharing second place. Nearly all poultry breeding is now in the hands of large organisations who breed what are termed hybrid stocks for broiler birds, layer birds and stock for specific purposes. Even turkey production, is mainly an intensive operation, with production still centred on the Christmas market.

Lastly, ducks and geese. Table ducklings are most grown in Norfolk and comprise almost half the national production. After intensive brooding, their period of infancy, they are fattened in large houses with grass runs attached, though these provide little more than shelter from the elements. Goslings are produced for the Christmas trade and are still raised to a table condition in the open, being the last of our feathered friends to live in natural conditions.

Before leaving the region mention should be made of the vital role of the livestock markets. Even Essex and Hertford, cheek by jowl with our capital city, maintain their markets, Essex having four and Hertford five, though two in the latter county handle only poultry these days. Modern legislation calls for very high standards of hygiene in all our markets. This means chiefly cleaning and disinfecting but has also resulted in improved construction and greatly improved the conditions under which animals are offered for sale.

Scotland

Our journey through Wales and England now being complete, we concluded our tour of the national farming scene by taking the road to Scotland. I have travelled up the West Coast as far as the near-wilderness north of Ullapool and marvelled at the contrasted rich farming of the East returning from Aberdeen to the capital city of Edinburgh. Certainly, two-thirds of the country consists of rough hill and mountain grazing. However, the Lowlands include some of the most productive areas in Britain.

The great crop of Scotland is undoubtedly timber with 10 per cent of the total land area being devoted to forests, but keeping to farming what better way than to record Scotland's role as principal producer of beef cattle in Britain, both pure-bred and cross-bred. Historically, this was the home of the finest oatmeal and until 1953 our family breakfasted on the produce of the Black Isle, sent direct by the sack from Dingwall. Certainly, the Scottish seed potato is without peer and principal growers in England depend upon it, secure in the knowledge that their crops will be free of disease.

The climate, contour pattern and soil determine the farming pattern as in Wales. Firstly, there is the great contrast in rainfall, comparable to Devon, some parts having over 2.54m annually. Sub-tropical gardens can be found in the sea lochs of the West coast, showing the humid conditions influenced by the North Atlantic Drift. Palm trees can grow as well as grass. The sunnier side of the country is drier and is ideal for mixed farming, at its best in Perthshire. Bleak, northerly winds can sweep in and blow the light soils and damage the crops in East Lothian.

Scotland has numerous islands and the three main groups have their own systems of farming. Crofting in Orkney and Shetland is the country's form of small-holding, often threatened, now more than ever, and yet still surviving. Here, beef is famed for its quality, the lower land also carrying good sheep. In the Outer Hebrides, farmers and crofters have improved the livestock capacity on their poor heathland while in the Inner Hebrides there is better quality land producing feed for livestock.

In the Western Isles and the West Coast, hand-shearing of sheep can still be seen and in the Highlands, the highest mountains in Britain, there are to be found very hardy sheep and cattle. Wheat is not important, except in the Lowlands, though the cool climate permits the growing of barley. The Southern Uplands, already referred to briefly, are heavily populated with sheep and grassland persists to the top of the hills.

The soil of the Lothians along the Firth of Forth. Berwick, Fife and parts of inland Perth (Tayside), the vale of Strathmore in

Our farming scene

Angus and the north-east coastal areas of the Moray Firth form, collectively, the real farming land. Mixed farming, fattening and arable areas, producing high quality crops including vegetables. The good farmland is not confined to the coastal lowlands but can be found in the river valleys, especially, the Spey, famous, too, for its fine salmon. Here is the home of beef cattle that take many prizes at the annual Smithfield Show. Travelling north, we find rich farmland around Inverness and in Ross and Cromarty, while in the moist south-west is a land of rich pastures, home of the handsome Ayrshire cattle and the main producing area for the country's milk, a quarter to be precise. Other milk producing areas are around Dumfries, Lanark and Aberdeen.

While a similarity exists between the farming of Scotland and Wales, there is in Scotland more output from a bigger arable acreage and more emphasis on beef rearing. I have attended the famous Perth Sale and sensed the world-wide role of the Aberdeen Angus, a supreme producer of early-maturing beef, popular in the United States, Argentine, South Africa, Australia and New Zealand. The high price Aberdeen Angus bulls fetch at the Perth Sale, 60,000 guineas was paid for a beast in the 1965 sales, confirms the breed's quality. The shaggy, long-horned Highland cattle are certainly as world-renowned as the Angus, while the Beef Shorthorn is its chief rival. The Galloway is another distinctive animal from Scotland and this list would be incomplete without mentioning the Luing, bred by the Cadzow brothers on the island of Luing, from Highland and Beef Shorthorn breeds. The Ayrshire, second-most popular dairy breed in Britain, deserves a second mention and to complete the cattle picture it should be noted that the Hereford has become established in the last 30 years.

Formerly, in Scotland, the chief function of sheep farming was to produce wethers for fattening into mutton in two or three years. Nowadays, the demand is for fat lamb—a lamb to give maximum lean meat for the butcher.

The two main breeds, Blackface and Cheviot have both made a great contribution to sheep breeding while the Border Leicester has been crossed and is in demand, both at home and overseas, as an 'improver'. Finally, mention must be made of the Shetland breed, whose wool is pulled when loose instead of, as normally, being shorn. This is used to produce the World-renowned Shetland knitwear.

We have wandered along way in this chapter and I hope I have made it clear to you the great variety of soils and weather conditions throughout the counties and the great variety of skills needed to farm these areas.

2 Becoming an agricultural apprentice

The first step

If you have decided that you would like to take a job on a farm, your school careers officer may have mentioned the Apprentice Scheme operated by a national body called The Agricultural Training Board. You will be concerned with it regionally and need to make contact with the local office either by letter or telephone. All the local offices are listed in 'Useful Addresses' at the back of this book.

Coming to grips: a preliminary course

Although the Apprenticeship Scheme caters for young people from the countryside, including the sons of farmers and farmworkers, there are many from the towns and suburbs who feel at a disadvantage because they lack any real knowledge of animals or the complicated equipment now used to cultivate the soil and harvest its produce. To meet this need there are colleges throughout the country which provide a short, preliminary course of 14 days instruction for school-leavers. Fortunately, the college staffs are in touch with training boards and can normally ensure that contact is made between the board's training instructor and those who attend this preliminary course. If you are accepted for a place at the course you will not be put to any expense apart from providing yourself with outdoor clothing for you will handle farm animals for the first time and tackle some field work with a tractor and implements.

Completing an application form

After you have phoned or written to the local office of the Training Board, a simple application form will be sent to you to complete, giving details of yourself, including hobbies, leisure interests and school career details, and exams passed or being taken. You will also need to give details of farming activities that specially appeal to you, for instance, sheep-rearing, tractor driving, milking, beef cattle, etc. Should horticulture be your interest, this cannot easily be fitted in with farm training. You should record 'horticulture' as your interest and not attempt to break it down into sections. Details of the structure of training in horticulture are given in chapter 6.

As soon as you have completed your application form, the local Training Board will make arrangements for you to be interviewed by the local Apprentice Committee or a representative group of its members drawn from the employers and workers sides. Also, a staff member from the college where you may have already taken the short preliminary course, and someone from the careers office, may also be present. Together, these people comprise an interviewing sub-committee, and as you may already be feeling some nervousness at the thought of an interview it will help you to overcome an attack of butterflies in the stomach if I tell you what to expect.

The Apprentice Interviewing Committee

This Committee will normally have a farmer 'in the chair', the remaining members will be sitting with him in a small, friendly group when you come for your interview. Hopefully, one of your parents will come along too, as this signifies you have support at home. However, a parent will not be there to hold your hand but, almost certainly, the committee will speak to Mum or Dad after they have talked with you.

I have purposely said 'talked with you' as you will all be discussing together the very important decision that you are planning to take. Also, you need to understand fully what is involved in apprentice training and avoid being a 'drop-out' from the scheme. Friendly interest will be shown in the details you have given about yourself on the application form and considerable care will be taken to ensure that you don't start off on the farming ladder as a square peg in a round hole. You will find that each of the members will want to talk with you about those interests in the industry they represent.

The farmer will almost certainly inquire into your ability to rise early and be willing to work extended hours when seasonal pressures require extra effort from all the staff—and you are hoping to become a member of that staff! The Trade Union member will certainly want to explain the wages and conditions prevailing in the industry. The member of the college staff, possibly the principal, will tell you about the day release studies you will undertake during your three years of training and satisfy himself about your ability to reach the college in time and return home at the end of your day's study.

An important person you may meet for the first time at the interview will be the Training Adviser, a full-time officer of the Training Board. He will be responsible for keeping in touch with you throughout your training, in a sense, monitoring your progress up the farming ladder. Additionally, he will talk to your training employer and satisfy himself that besides being involved in the daily work of the farm you are grasping its rudiments from the outset and throughout your training period.

When you have asked any questions yourself and Mum and Dad have joined in the gathering to satisfy themselves about those little human considerations that *all* parents regard as important, you will be told the decision of the committee. If you are told you will be advised of the committee's decision in writing, but it is usually possible to form a clear idea as to whether you've got through or not before leaving the room.

The training employer may or may not, be a reality at the time of your interview—I will explain. Normally, the Training Board expect an apprentice-to-be to scout around and find a farmer or grower who is prepared to offer him employment as an apprentice. There is a variety of tactics to be employed in the quest, all of them effective, but one which never fails to winkle out an opportunity, if one exists, is that of making a personal call on a farm near to your home or in a place where a tradition of holiday making has established friendly relations with local farmers.

This does not mean other methods cannot or should not be employed. It merely signifies that farmers still tend to be individuals who attach importance to personal relationships within an industry where employer and employee rub against each other and follow identical paths work-wise in a way seldom found in town or industrial work.

Now back to these other methods of finding a training employer. There are occasionally vacancies advertised in the local paper of any county town and many country towns but vacancies are not always notified at Job Centres. The National Farmer's Union have a journal circulating in your county and it is a very good idea to offer your services in its advertisement columns—normally no charge is asked for by the Union's Regional Information Officer to whom applications should be made (see 'Useful Addresses').

The urban would-be-apprentice may appear to be at a disadvantage compared to his country cousin, seeming to have unlimited opportunities in the field of personal contacts. While it must be admitted that the closeness of village communities can often lead to doors being opened, this is far from the whole story. More and more farmers are becoming aware of the dedication and enthusiasm that the town boy or girl can bring with him to the farm when starting work.

Should all else fail, and you come to the interviewing committee lacking a training employer you will find both the members and the training adviser sympathetic to your needs but it has to be emphasised that the Training Board *do not* accept responsibility for providing you with a job.

Many apprentices-to-be ask, *'How will my school exam results affect my training?'* Almost certainly, the gaining of good GCE or O level passes and an aptitude to learn the theoretical as well as the practical side during your apprenticeship will lead to the training adviser and the college of agriculture recommending that your third and final year of apprenticeship training be devoted to full-time college study. The cost of this, as with day release training is covered by a grant. Becoming an apprentice is not 'returning to school'. Never neglect your day release.

Day One as an apprentice

We can now assume that you have jumped the interviewing hurdle and landed successfully in your first job. The final day at school or school holiday approaches and your training employer has tied up with you the day your new job will start. Your holiday arrangements will normally be accepted by him but, if possible, enjoy these before the appointed day. As this approaches thought must be given to the very important point: *What shall I wear?* Farming is not a uniformed occupation though when it was a mainly male occupation the smock was standard, until well into the last century. My own first contacts with the work-force almost 50 years ago were at a time when uniformity of dress revealed itself in the number of variety of waistcoats worn and these were *not* discarded one by one as the day wore on.

Today, the important point is the seasons and the nature of the duties you will be called upon to carry out. The modern tractor is fitted with a safety cab capable of insulating you against the weather but the need to be involved in adjustments and running repairs, both to the machine and the linked implements, dictates the wearing of a strong boiler suit and donkey or similar jacket, the latter being discarded in the cab and certainly not needed in the summer months.

Stout footwear is a 'must' at all times. even when one's personal feelings jib at the idea. For much of the year rubber boots or 'wellies' are needed—always when working inside buildings where stock of any type are housed. Apprenticeship opportunities occur within the specialised field of intensive poultry production and rearing and also in egg production. Here protective clothing in the form of overalls, 'wellies', and headgear is a *must*, too. being normally supplied by the employer.

You are bound to ask, 'do I supply my own clothing or does the training employer?' the answer is straightforward but there are snags. When any clothing item is required to ensure compliance with the Health and Safety at Work provisions, the employer is legally obliged to provide you with what you need. Because health and safety at work is so important to you a later chapter in the book explains how it will affect you.

Oh yes, you are asking but what about these snags in connection with clothing. Well, it revolves around the nature of the particular task you might be required to perform and whether the task has a health and safety aspect. At the time of writing this book a common sense approach applies over this matter but it is possible that if everyone concerned does not make an extra effort to improve the safety record in agriculture the law may be given additional teeth. This could affect the whole matter of 'what shall I wear' and who provides what.

Lastly, keeping out the elements, rain, hail, snow, gale force winds, hot sunshine and dust. Coming from the south east of the country you would not expect the extremes of weather that are a normal, seasonal hazard in other regions of Britain. Nevertheless a full length, waterproof coat should always be with you if the job to do is outside and fine weather unlikely—it's better to be safe than sorry—and nothing depresses the newcomer to farming more than getting soaked to the skin.

3 Entering agricultural college and gaining practical experience

Unlike many jobs, farming and horticulture require of their entire work-force some practical experience. You can work on the farm or on a nursery without training and I will say something later about going straight from school to a job on the land. Nevertheless, if you have a career ambition and mean to go places, either in the industry or within its 'fringe' activities, you come back to the need to obtain qualifications either through the Apprenticeship Scheme or by going to agricultural college.

Having said that, you will find as you read this book, that you *can* get into farming without first taking examinations. In fact, the majority of entrants do just that, but realising, as they grow older, the place of qualifications, they train and study alongside their daily work. This, too, will be dealt with later on. I am quite certain that much of the objection to becoming an apprentice in either farming or horticulture stems from the desire of a school-leaver to have done with study, put the classroom behind him and get launched on the adult activity of 'going to work and having money of your own' which has been earned by daily labour.

Becoming an apprentice links you up with the Agricultural Training Board. Going to college keeps you well and truly linked up with the educational system of this country. If your school record suggests you have special ability that can be used in agriculture and horticulture on a managerial or administrative level this chapter is specially for you. Your county agricultural college provides the course of study you will need to give you access to this level of employment and the degree of academic skill you show will determine your point of entry.

The education in agriculture in the United Kingdom is at three academic levels:

1 *University,* where entry requirements are normally two science A levels, the degree course chosen taking three or four years to complete.

2 *Named colleges,* eg Seale-Hayne in Devon, the 'Royal' at Cirencester in Gloucester, Harper Adams in Shropshire and Shuttleworth at Old Warden in Bedfordshire. These provide either Higher National Diplomas or Ordinary National Diplomas. Entry requirements for the first named include one science A level; with four O levels or four Grade One or CSE passes, which include two

science subjects for the Ordinary National Diploma entry. Both courses are normally three year sandwich courses. This means you do a year at college, a year of practical work and a final year back at college.
3 *County colleges,* sometimes still called by their original title of Institute. To enter these, your school examination results can be modest. In fact, they are minimal compared to the requirements for other forms of agricultural education but the intending student must be adjudged capable of benefitting from some academic, as well as practical, training before being accepted. Courses taken are normally the National Certificate in Agriculture which lasts one year, with some county colleges doing the OND courses as taken at the named colleges. Many county colleges have developed special residential courses related to the needs of the region where they are situated.

Such needs would include amenity landscaping, where closeness to large built up areas requires such a course, and, you will recall, the important one-day release for apprentices takes place at the county college.

A minimum of a year's practical experience before all forms of residential training are taken is normally required, or is a condition of entry, and at the outset you would be well advised to approach a likely employer in a mood of willingness to be really useful on his farm and be willing to turn your hand to anything that crops up in the day-to-day operation. This could take the form of helping to effect a repair to the milking equipment as soon as you arrive at work, rushing into town for a vital piece of equipment, or even assisting in the workshop or moving farmstock with which you have not had time to familiarise yourself.

What practical work will I be expected to carry out?

Farms do not fall into exact categories neither is there a common pattern of operation related to their hectare size. It can be regarded as certain that any farm requiring labour to assist the owner, or the owner and family, will provide the opportunity to do your year's practical work.

The routine for obtaining a job which I outlined in the last chapter will certainly be worth following and your local National Farmer's Union office can be turned to for help and ideas. Having also followed the guidance I have given you about what to wear you should be ready for anything!

As to the work you will be expected to do during your

pre-college year, your personal approach, summed up in the last paragraph is more important than the varied tasks that will, almost certainly, come your way. It is vitally important to start off in the right frame of mind.

If you are to help with stock, always wait to be told about the animal concerned, and what you are required to do in connection with its rearing and well-being. With the best will in the world it will be difficult for you to grasp the amount of working capital that can be tied up on the four legs of a cow, heifer, bullock or the bull connected with an average dairy farm. To a lesser extent, the same applies to the sow and litter of a pig unit—as well as the boar who can prove to be quite a formidable customer. Sheep, no less, whether they be ewes, lambs or the ram of the flock, all constitute a lion's share of the holding you may find yourself working on for the year.

On a mixed farm you will be called upon to help with soil husbandry tasks that will call for a measure of skill from the outset, beginning with the tractor which is now a complicated item of tackle. More important than driving it is knowing how to use it economically, as the price of fuel is an ever-increasing headache to the farmer. It is worth mentioning that the careful use of a tractor and for that matter, your own car or motor-cycle, can help to restore this country's economy to a healthy state. In fact, if we could all regard fuel economy as fun, our national fortunes would soon take a turn for the better!

Driving the tractor requires skill in maintenance as well as operation and you will soon learn that on the farm job demarcation is unknown so you want to know how to keep your tractor rolling whether ploughing, cultivating, spreading fertilizer, crop spraying or even carrying out the many fork lift operations that have, to a large extent, superseded the hard grind of manual labour that would have been your lot a few decades ago.

You will be called upon to help with the combine harvesting, baling and drying of the wheat, barley or oats, possibly showing an early skill with these giant machines that can represent five figure expenditure to the farmer. Then there may be potato growing, sugar beet, oil seed rape, grass and forage and even pea vining if the enterprise you work in is big enough. But never forget, in farming, as in most things, small can be beautiful.

There remains *the* question still uppermost in your mind, despite this explanation of how to get to an agricultural college and what you are likely to encounter in your year of practical work (remember, one year is the minimum). The question undoubtedly is, 'What is my local college like?' Only one reply can be given to this and to the further questions that naturally arise from it, related to curriculum, facilities, accommodation, meals and leisure and hobby opportunities.

The reply to your question/questions can only be given by the Principal of your own county college. In the first place you must write to him so that in a resulting interview he can assess whether you meet the requirements given earlier in this chapter under the heading: 3 *County colleges.* In the course of the interview all the obvious questions are certain to be dealt with and an opportunity given to look around the college and its farm buildings. You will find that the college will reflect the farming pattern of the region in which it is situated and help to bring alive what was written in the first chapter of this book. Should horticulture, poultry rearing or any other specialised field be important to the region's economy you will find the curriculum will embrace them. For instance, Dorset College of Agriculture tells intending students, 'We have purposely set out...to cater almost wholly for the needs of the rural county of Dorset'.

Finally, what happens to the students who have successfully completed a course at agricultural college or obtained a degree and passed out successfully in their university courses? Here again, you will find the principal or members of the staff will have a record of where the students have gone and what they are doing in terms of career achievement.

Saddleback sow and litter: an old English breed good for crossing for bacon and pork

4 Working on the farm, outside the Apprenticeship Scheme

In the previous chapter I have explained something of the conditions to be found in pre-college training but it would be safe to assume that over four out of ten entrants go straight to a place of work without any formal training or further education. Should this course of action appeal to you this chapter will mention possible pitfalls without in any way pouring cold water on what, after all, can be the spirit of adventure coming out in you.

From talking to young people who want to leave school at the earliest opportunity I have come to accept that when their minds are made up about going out to work the best thing to do is to let them get the bit between their teeth as soon as possible. Although the majority of farmers, the people to give you a job, are now showing a preference for apprentices this need not discourage you from seeking out an employer who will be willing to take you straight from school.

The first advice I give you is *don't* leave looking for a job until your last term is approaching at school. You will be more likely to find what you are looking for if, in your holidays, you have made an effort to get involved with life on a farm by offering your services—and it doesn't matter what the job! Many farmers are organised so that there are a variety of jobs about the place to be done but they are not sufficient to justify the employment of a full-time worker. This could result in you getting used to handling animal feeding stuffs, cleaning up and even handling stock.

One of my favourite farming books is by Briscoe Moore, called *From Forest to Farm,* published by Pelham Books eleven years ago; you may be lucky enough to find a copy in your local library. Some advice he gives I would like to pass on to you. He emphasises the need to have a knowledge of chemistry in an age where such a variety of stock remedies, weed hormones and sprays are used. He also considers that a young person thinking of farming as a career should be a carpenter and a mechanic so if you've developed skill at these while at school tell any would-be-employer you approach for that opening on the land.

As you read through this book you will find more and more references to the sort of things that go on in most farms. It is easier to deal with them in this way than to be greeted with a chapter totally devoted to a list of farm chores and duties, or the great variety of tools and implements you will come across when attempting to perform them.

A holiday job will enable you to see something of fencing, farm plumbing, drain-laying, building repair and may be that mystic operation of hanging a gate. You will certainly be at an advantage if you have a farm or country background as there is still a good deal of sense in the old saying, 'like father like son'—most potential employers take it as read that a farmworker's son who is keen to continue the family tradition will have a feel for the job.

Should you decide to enter farming in this way you will not be at a disadvantage as far as your starting wage is concerned but it is only right to advise you that to maintain your position as the years go by you will need to take advantage of the *Craft Skills Training Scheme* and obtain a certificate which will enable you to claim a better wage where you work.

This Scheme has proved successful and become a permanent method of craft training in the industry. Operating alongside the Apprenticeship Scheme it will cater for you as a worker who has opted to enter farming outside apprenticeship. Additionally, it will include some skills which are not covered by the Apprenticeship Scheme and include horticulture as well as farming.

Training arrangements are made to suit you and your employer. The training is of a practical nature, like the work you will already be doing. The practical courses last about one or two days either 'on the job' or at a suitable centre away from your place of work. In addition, you will be shown how to do various skilled jobs in the course of your actual work. All this will lead up to an expectation that you will train for a number of tests needed to qualify as a craftsman.

At this point you will approach the National Proficiency Tests Council, which organises tests throughout the country, normally close to a worker's place of employment. The training itself can be arranged to suit you and enable you to practise your new found skills until they become second nature to you. Those starting from school usually need three years to complete all the tests but this can be less if you find yourself working with intensively kept livestock. If it suits you and your employer to take more than three years, this is up to you.

Hopefully, you will not feel that going straight on to a farm from school is a practical proposition provided you can: (i) find a suitable employer, (ii) once you settle down to working with him agree to take part in the craft skills training scheme and become a farm craftsman by the time you are 18 or 19.

Working on the farm outside the Apprenticeship Scheme

There remains the matter of cost, something that nowadays is described as a 'variable' although, as in most things, it would be safe to reckon that the figures I quote are more likely to go up than down! You will be called upon to pay between £2 and £5 per test activity while the Agricultural Training Board pays the local test service £20 towards the cost of providing each candidate's qualifying test. Additionally, the Board pays the instructors and meets many expenses of the course organisers and, very importantly, it keeps in touch with you and your employer to help your training along. You are given free of charge:

- a simple log book to record the training and testing you do;
- details of the proficiency tests you have to take;
- training guides that remind you how to do jobs after instruction has been completed (no need to write any notes!).

The Board also pays a grant to your employer when you pass tests. This is to encourage him to join the scheme in the first place, and to help make sure you get the training and experience needed to pass the tests and qualify in a skill.

Contract or self-employed workers

In chapter 12 reference is made to those who have an ambition to work for themselves rather than for someone else. Although *not* the way to start a life on the land, there is a case to be made out for becoming a contract or self-employed worker once a measure of skill has been learned.

Contract work means striking a bargain with a second party to undertake work to supply a product or service at a price agreed by both parties, still common in many fields of crop growing and in forestry and garden landscaping. Because of the strong element of risk, this is not a field of activity to be recommended to those starting their career on the land.

Self-employed workers undertake to perform a variety of rural jobs for a set rate and pay their own National Insurance contribution. The employer does not have a National Insurance liability in this situation and would not normally accept a training role as they would towards you when starting your life on the land. Self-employed persons have difficulty in enjoying the protection afforded by the Employment Protection (Consolidation) Act, 1978, explained in chapter 10. Here again, this is something to look into when you have a secure foothold on the land, but not when starting your career.

5 National Proficiency Tests

In previous chapters we have considered ways of becoming a farm worker. What is comon to them is that they provide the opportunity to become qualified as a craftsman, and receive the premium wage in excess of the ordinary worker. For many years there has existed in this country an organisation called the National Proficiency Tests Council. Operating on a county basis it seeks to maintain craft skills in both farming and horticulture.

At the same time it has to recognise that although many skills remain unchanged for generations more and more require to be up-dated. It has to ensure employers and workers, with the skill and interest to serve as examiners are themselves aware of the changes and of what are termed 'new techniques'.

In the county where you find work there will be a Proficiency Test Committee. Its secretary is usually located at the local agricultural college, a place where, sooner or later, you will establish contact as you seek to join with others in gaining skill and you should have no difficulty making known your wish to be tested in farm skills, once you have reached the required standard. Again I would stress the importance of requiring skills; they will unlock so many doors to you. One difference between the craftsman and the unskilled worker is, whereas the craftsman finds himself caught up in all that's going on in the local farming scene, the unskilled worker, because of his isolation, tends to live the life of a 'loner', with job interest ceasing when the day's work ends.

A summary of the contents of the agricultural tests coming under the wing of the National Proficiency Tests Council will remind you of the varied nature of farming operations.

Milk Production
Milking and Dairy Hygiene; Dairy Cattle Stock Tasks.

Beef Production
Beef Cattle Stock Tasks.

Sheep Production
Sheep Stock Tasks; Sheep Shearing; Sheep Dog Handling.

Pig Production
Pig Stock Tasks.

National Proficiency Tests

Tractor Driving
Tractor Test.

Mechanised Operations
Tractor Test; Machine Maintenance; Cultivating; Sowing/Planting; Spreading; Spraying; Harvesting; Fork lift truck or tractor; Crop Processing; Boundary Maintenance; Specified Welding.

Farm Maintenance
Fences or gates; Hedges or walls; Drains or Ditches; Power saws; Farm Joinery and Maintenance of Woodwork; Care and Maintenance of Small Hand Tools; Farm Brickwork or Concrete Blockwork; Concreting; Farm Water Supply, Repair and Maintenance.

Poultry Production—Turkeys
General Maintenance; Operation of Equipment; Handling of Stock; Routine Husbandry; Breeding and Egg Production; Slaughter and Preparation for Direct Sale.

Poultry Production—Broilers/Capons (these names are given to chicken intensively reared for eating, what the trade calls 'the table')
General Maintenance; Operation of Equipment; Handling of Stock; Routine Husbandry; Breeding and Egg Production; Slaughter and Preparation for Direct Sale.

Poultry—Eggs
As for Poultry Production.

6 Careers in horticulture

Varied tasks and opportunities

The important field of nursery, vegetable growing, landscape gardening, fruit growing and the production of watercress form the considerable industry which, grouped together, is known as horticulture. Whereas farming is common to all regions of our country, horticulture tends to group itself in certain selected areas, often conspicuous by acres of glasshouses, neat rows of fruit trees or bushes and the varying greens that the great variety of vegetable grown in this country present to the passer-by. If this country is to possess a horticultural industry able to compete with the highly organised, and often subsidised, vegetable, fruit and flower growing activities of our partners in the European Economic Community, we need a skilled people at all levels.

Certainly, if you are to enter the work-force that produces these essential foods you will need to possess dexterity with your hands, great patience, for there can be much repetitive work, a good memory, and willingness to accept that it can take the best part of a lifetime to grasp all that should be known in the branch of horticulture you decide to enter. There may be someone who can master fruit-growing in its amazing variety, as well as bulb growing and flowers and vegetables, but you are unlikely to meet such a person. Hence you will need to decide the branch likely to appeal to you and, just as importantly, with a work opportunity existing where you are living or where you are prepared to go.

National Proficiency Tests – Horticulture

The following summary will give you some idea of the many and varied jobs involved in horticulture from which you will soon see whether this is the career for you.

Glasshouse Crops Production
Preparing border soil; Sterilising soil; Preparing compost; Applying water, liquid and solid feed; Planting (in border or container); Controlling glasshouse environment; Supporting, training and trimming plants; Glasshouse clearing and cleaning; Harvesting; Soil block making and mechanical planting.

Nursery Stock Production
Glasshouse/Case/Frame propagating; Preparing growing medium; Planting or potting; Pruning, staking and training; Field lifting; Plant identification.

Outdoor Vegetable Production
Preparing propagating area; Soil block making; Planting; Singleing and hoeing; Harvesting; Transporting empty and full produce containers; Loading road transport.

In addition, the following are common to the sections listed above:
Propagating; Grading and packing; Tractor driving; Pedestrian Controlled Rotary Cultivator; Recognition of, and control of, pests and diseases; Mechanical potting (*not* in outdoor vegetable production); Irrigation (*not* in glasshouse crops production).

Landscape Practice
Planting; Turfing; Seed sowing; Rock work; Operation of machinery; Walling; Paving; Concreting; Gravel surfaces; Operation of pedestrian controlled machinery; Pruning; Lawns; Planted areas; Herbicides.

In addition to these horticultural tests there are three tests in *Bulb, Corm and Bulb Flower Production* and 14 in *Watercress Production.* Although few candidates present themselves for testing in these two fields they are both of considerable importance; the first-named within the pot plant trade and the second because, although regarded as 'old fashioned' it is, in fact, a nutritious and increasingly appreciated salad food.

Living accommodation

Some large nurseries in this country provide hostel accommodation for students and apprentices but they are few and far between. Even rarer, particularly in the South-East of England, can lodgings away from home be found. Here again, your local National Farmers Union will be the best source of information both for work opportunities and the likelihood of accommodation being available. The NFU horticultural section, as well as being a powerful pressure group for growers and orchardists, concerns itself with the local needs for labour and the housing of workers.

Other factors when seeking work

Although the wages and conditions in horticulture resemble closely those in agriculture, including a standard working week of 40 hours, the efficient running of where you seek a job may call for overtime working. Market needs, watering and other essential tasks can rule out the likelihood of an eight to five pattern operating, though this can depend to a large extent on the time of the year.

Clothing advice you will have already digested in Chapter two, so it only remains to emphasise where horticulture varies from farming. Humidity and moisture pose special problems when working under glass, while many jobs in fruit growing call for a clothing pattern that varies in the course of the day as the temperature rises only to drop as the afternoon wears on.

Lastly, those things that seem commonplace but may not have crossed you mind:

Tall people will find working under glass rather cramping!

Although good eyesight helps, with a pair of spectacles and the ability to develop powers of close observation you should do as well as a 'Hawkeye'.

Never consider having you main meal of the day during the lunch break, better to experience an aching void in the later afternoon than the discomfort a full stomach brings should your work call for bending, reaching, stooping — as it so easily can in most horticultural work.

The same applies to alcohol, leave it alone until the day's work is done.

Vegetable growing

Nothing will appeal to you more, if you are inclined towards horticulture, than the great variety of vegetables that can be grown in this country. Fascinating too is the history of their origins. Many vegetables were known to the Egyptians, Romans and Greeks—and it is to the Romans that we owe the greatest debt, not only for a great variety we now eat but for their methods of cultivation which still form the basis of our modern practice.

History, whether in farming or horticulture, is not a laid-down essential but I am of the view that if you are serious about making a career in either you will delve in the library and begin to find out more.

Fruit growing

Should you feel drawn to orchard work you will certainly find out, sooner or later, that the industry is under threat from overseas imports, but don't be deflected from a life full of variety and fascination, planting, ring-barking, ringing, bark slitting, budding, grafting—a science in itself, manuring, both natural and artificial, pruning, branch bending, thinning, renovating, grubbing, de-horning, spraying, picking and storing—all these activities can come your way in apple growing alone.

There are old varieties of apples with bawdy names which I have heard about but never actually seen written. Only 37 years ago the distinguished fruit grower, Raymond Bush, listed 65 varieties of cherry, 77 plums, with an additional 12 known as incompatible, 70 apples and 33 pears. Then it is likely that a fascinating list could be made of such fruits as nectarines, quinces, peaches, apricots, mulberries and medlars growing both on trees and bushes.

However, you are unlikely to come across all of these if you embark on a career in fruit. You might easily become involved in soft fruit growing, which includes black and red currants, gooseberries, loganberries, boysenberries, raspberries, and strawberries. What they all have in common is their ability to contribute tasty, nutritious and very pleasant tasting additions to our diet.

Mushroom growing

Mushrooms, which started as a field crop, were collected from high summer onwards and were a much sort after addition for a flavoursome meal, have now become a large scale, specialized industry with few openings career-wise. Most of the intensified production work, in rows and rows of houses from which light has been excluded, is carried out by unskilled or semi-skilled labour and is, very much, a conveyor belt operation.

Flower growing

I'm sure you are beginning to realise how much there is in horticulture, but there are branches offering job opportunities still to be examined, in particular flowers. A very concise leaflet on 'Careers in Horticulture', published by the Agricultural Training Board, mentions over 400 hectares of glass devoted to flowers including carnations, roses, fresias and pot plants. The total area is immaterial but it includes large rose gardens and the fields of glorious flowers, particularly in *Lincolnshire* with early daffodils in the *Scilly Isles* and *Cornwall.*

London, and for that matter all our principal cities, have always provided a large market for the flower trade whether cut, potted or as plants and bulbs. In recent years the first two have grown enormously and it is on the outskirts of our cities that the growers, both large and small, can be found. The 'House of Rochford' springs to mind. Situated some 14 miles north of London this conspicuous area of glasshouses in the parish of Turnford has a splendid apprentice scheme, hostel accommodation for students from home and overseas and is in the forefront of places offering careers to the beginner.

Intensive vegetable growing

Earlier in this chapter I mentioned our heritage of vegetable and fruit growing. However, to prevent us reaching for the stars, I have also strongly hinted at the economics of present day agriculture which can be summed up in the word 'commercial'. In the simplest language this means you must cater for market trends and bend to the forces these trends represent.

Nowadays the growing of vegetables is intensive, with the machine more and more taking over as it has in farming. For over 30 years, continuous research has been carried on at the National Institute of Agricultural Engineering, Silsoe, Bedfordshire, to perfect and produce machinery and equipment to replace hand labour in every branch of vegetable production. In addition, institutes and private companies have put a great deal of know-how and experience into achieving the same end.

With over three million tonnes of vegetables being produced yearly—and this figure does not include the potato!—the large scale growing of vegetables has become part of normal farming practice so you will see why horticulture and agriculture have to be linked when talking about 'Working on the Land'. In addition to these millions of tonnes of vegetables, there are 200 hectares producing glasshouse crops, tomatoes, lettuce, courgettes, cucumbers, peppers, etc.

Packhouses

Each year sees an increase in the quantities of peas, beans, carrots, and Brussels sprouts grown for the freezing industry. To compete for the favour of the housewife, the harvest of vegetables is more and more being handled by packhouses to reach the shops and supermarkets as a packaged, convenience food, merely requiring a price tag to reflect the state of the market.

Packhouses can be run by private companies or linked with the actual crop production as in the case of the Land Settlement Association, a co-operative group which came into existence during the 'depression' when the unemployed were offered a worth-while life on the land. In a later chapter more will be said about this organisation. Now I am thinking of one of their groupings of small-holdings at Chawston in Bedfordshire. A centrally situated packhouse receives the products of these holdings daily and, using modern equipment and the back up of a skilled, experienced work force ensures the arrival of vegetables in the London market early the next morning. Ten miles south, down the A1, two miles east of the old market town of Biggleswade, can be found the packhouse of Beds Growers, another co-operative enterprise, which receives the market garden crops from the local growers, packing and marketing promptly in much the same way.

Should you live in any of the country's vegetable growing areas similar set-ups can be found, seeking staff from time to time. Although the English fruit trade has been the object of much criticism regarding its grading and packaging, if you live south of a line drawn from the West Midlands to The Wash, well-organised and efficient packing sheds can be found on the majority of orchard holdings.

Soft fruit is certainly a crop of the present day. Strawberries, raspberries, blackcurrants, gooseberries and blackberries are often produced on holdings similar to those described above. However, their distribution is much more widespread and includes two areas of Scotland, the east central part of that country and the Clyde valley so work opportunities exist there, too.

New methods and techniques, similar to those already described, are being used to develop soft fruit production. Many amazing machines are being used both for cultivation and at the time of harvest. The actual tasks you would be called upon to perform are similar to general orchard work and, in particular, the careful and regular cutting of grass is essential as, left untended, it is a great robber of that essential plant food nitrogen—no soil element will be more important to you when you start working on the land.

An ATB Leaflet, *Careers in Horticulture—Working in tree and soft fruit production* emphasises the need to have a knowledge of horticultural chemicals and specialised machinery. In addition, and something I've not mentioned before in this chapter, the ability to keep accurate records. Over and over again it must be hammered home that the Landworker of the eighties, irrespective of his role in the industry, needs to organize his work, and keeping records is essential.

Amenity horticulture and landscaping

As modern architecture and town planning replace the soft lines and gentle proportions of our towns and cities, amenity horticulture and landscaping take on increasing importance. We are a nation of country lovers, in the main, and nothing eases the eyesore of stark concrete and macadam roads than patches of green and the splash of colour well-chosen shrubs and flowers can give to our townscapes.

The role of Forestry will be dealt with in detail in chapter 6. When considering work 'on or about the land' we need to realise how important trees, shrubs and bushes are to a community. If you want to spend your life with them amenity horticulture and landscaping could be the career for you without going into the forest. Two main avenues of work exist, with Government and Local Authority, or with a private firm of landscape gardeners or contractors. Also, there is one other important but limited field of job opportunity. The National Trust who would be offended if the impression was given that they are linked to the Government or the private sector.

The National Trust is a unique British Institution entirely dependent on the support of the people of this country aided by concerned overseas membership. You may be one of the millions of people who visit the great houses, monuments, gardens and reserves which they own throughout the land—The National Trust needs to maintain its houses to a very high standard; furnishings, decoration, art treasures and floor coverings must all combine to give an impression of perfection. Outside, the same standards must operate if the continued support of the general public is to be assured. In the gardens, parklands, woodlands and farms, attaining this standard provides horticultural and agricultural job opportunities, and some 80,000 men and women are employed; many engaged in *Amenity horticulture and landscape gardening.*

The skills involved in working in this field are best learned by *serving an apprenticeship*. This will give you an understanding of soils, cultivations, drainage, fertilisers and the various types of grasses; an ability to use and maintain a wide range of machinery;

a knowledge of the use and application of building materials for pathways, walls and fences, and of the use of simple levelling and chain survey techniques; an understanding of plans and drawings; of how to use chemicals to eradicate or control pests, diseases and weed, and how to maintain landscape areas, and a knowledge of a wide range of plants, trees and shrubs.

Prospects

Whether you are looking for an opening with Government, Local Authority, or a private contractor you will want to have some idea of what the career prospects are after you have achieved craftsman status. The next step would be to take charge of small groups, usually two or three, as a *Chargehand.* Then going up the ladder you could take charge of a site and those employed on it. This would involve you in keeping site records. You would be the *Foreman* and would almost certainly be expected to drive a vehicle. The final rung in the ladder would be as a *Manager*, with responsibility for the supervision of the foreman and organisation of the sites. Work records, programmes and the carrying out of contracts would also be included in your duties.

Having outlined the career opportunities let me express a personal view about job prospects in this and every job related to the land. Your common sense should tell you that with each step up the promotion ladder the number of staff required lessens. Good luck to you in seeking whatever ambition you have but never embark on a job assuming you will be the selected candidate. If, for some reason, opportunity passes you by, you can be assured of enjoying comradeship, satisfaction and, above all else, happiness with the passing of the years.

Pay and hours in Horticulture are normally those operating in agriculture, notes on which are given in chapter 10. However, it should be pointed out that a different wages and conditions structure might operate if you are employed by a Local Authority.

What the future holds for horticulture

Despite the problems of foreign competition to our growers, the cuts in public spending affecting landscaping in our towns and cities and the spiralling costs of everything used in the industry, this is a growth industry. There is a need for every scrap of food that can be produced in the world and the way will be found to overcome the distribution problems. In a sane world there cannot be starvation for some with gluts and surpluses for others. To ensure a balanced, healthy diet for mankind we need all the vegetables and fruit our diminishing, fertile soil can produce. As man does not live by bread alone, we certainly require all the beauty and cheer that flowers can bring and the shade and colour shrubs and trees can provide in our towns and cities.

7 The future size of the agriculture labour force

When this book was in the planning stage few people would have predicted the onset of a serious economic depression. At the moment there is no evidence to suggest that agriculture will be affected as seriously as manufacturing industries. Cuts in capital expenditure will affect land and rural-based jobs. What the newcomer to farming may have to face is running into competition with job-seekers who, in the past few years, you would not have seen when job-selection was taking place.

The purpose of this chapter is to provide a sounding-out section for those who feel life on the land is very much for them but want to look ahead a little and seek to find out the pattern or working prospects in as far as they can be gauged. The challenge of farming as a career remains constant in good and bad times and, come what may, this country and the World will always need to be fed.

I would emphasise again how the nature of food growing has been changed by the yearly gobbling-up of the nation's land for other purposes. While at the same time, thousands of hectares within our urban and town boundaries become twilight or derelict zones, which are often ideally suited for a return to their original role of food producing; more importantly they are sited close to a ready market for the products they could so easily grow.

At the time of writing it is not easy to get exact figures, neither would these prove much, anyway. There are about 7,000 farm managers, about 213,000 full-time workers and 80,000 seasonal or casual workers. Although it is not easy to obtain a farm, men and women become farmers every year and there must be nearly 200,000 of them in the British Isles. I have said something already about the job opportunities on the fringe of agriculture and certainly these would add hundreds of thousands to the grand total.

What is Government policy?

Although I hesitate to encourage anyone to depend on support from government, we should look at a Whitepaper as such documents are called, considered by Parliament in 1979. Called 'Farming and the Nation' it said many optimistic things, supported by the use of figures and statistics that are always an important part of such documents. There was a change of government in 1979, but no change has occurred to alter the substance of the things said as they affect you and your hopes to work on the land.

Although the language used is rather 'heavier' than elsewhere in this book it seems right to quote this document and start with a statement of the fifth page: 'The rate of decline in agricultural manpower has slowed down and this trend is expected to continue. The Government recognises the important contribution of education and training to the efficiency of the industry and will continue to encourage the development of services appropriate to the needs of the farm work force. Greater skill means more productivity and higher pay, and the Government looks to see both in the years ahead.'

These sentiments are spelled out on later pages, but, above all else, the message is 'if you are willing to learn, whatever your ambitions are, you will find support to do so and, more importantly, your rewards will be greater' and, to quote from the text of the Whitepaper: 'There is now an onus upon farmers and farmworkers alike to ensure that full advantage is taken of these opportunities' ie to be trained.

Before we leave 'Farming and the Nation', reference should be made to the optimism expressed about farm produce in the future. Estimates, as well as hopes, about increasing production of farm commodities imply the need for more labour and must affect the size of the work force. In particular, high hopes are expressed about the rate of increase in horticultural production, which, to a greater extent than farming, is labour intensive.

Most studies in recent years on future employment prospects on the land, have been of a local or regional nature, in particular, the agricultural economics units at many of our Universities have delved into the subject. A paper by Vic Beynon of Exeter University called, 'The Economic Importance of Devon Agriculture' spells out that farming means work—perhaps now is the time to blazon this message sky high and in capital letters! Certainly, with agriculture providing more jobs than anything else in the county of Devon, there must be lessons to be learnt for the rest of the country.

In Britain we have a National Economic Office. You may well ask, what has that got to do with my prospects of finding a job in the future. Actually, under its umbrella there is a department called the Agricultural Economic Development Committee. Although *not* a government department or involved in putting into effect government decisions, it is worth noting that a report it published says things about jobs on the land, and training for them, later to appear in the government Whitepaper mentioned earlier.

Called: *'Agriculture into the 1980's—Manpower',* the report is the findings of a group of people from all sides of the industry; farmers' representatives, agricultural training people, trade union leaders and country landowners, under the chairmanship of Mr Reg Bottini CBE, for many years the General Secretary of the National Union of Agricultural and Allied Workers. Together, they considered the future labour needs of the industry in great detail. Reading it again, after almost three years, one notes that the decline in the numbers of those working on the land would lessen from 2 per cent to one and a half per cent until the early 'eighties'.

You may feel this a rather trivial matter but it has to be considered in the light of the number employed in the industry given at the beginning of this chapter. It must also be considered in the light of declared government policy to grow more food from our own resources. Actually, a government Whitepaper called, 'Food from our own Resources', published in 1975, and up-dated from time to time, was used as the background for the group's report.

One useful job done by the group was to study the reasons for the fall in the labour force over the years, something important to you in coming to a decision about working on the land, as knowing what the future needs may be, is vital. When writing about the latter there must be an element of guesswork, as in all planning, whereas in telling you about why people have left the land, I am dealing with facts. The fact that stands out above all others is that there are now more full-time farmers than full-time hired workers, while full-time hired workers have become more concentrated on the larger farms, as described in 'Our Farming Scene' at the beginning of this book.

Only about a third of the full-time holdings in Britain hire full-time workers outside the family, while less than a fifth hire two or more. The report maintains that the growth of family farming is not confined to what are called 'smaller units'. Larger holdings normally employing full-time workers are more and more getting by on using the family for all the chores, aided in times of seasonal pressure by what are termed casual, contract or part-time workers. I will explain about these groups of workers in a later chapter as one of the three categories may appeal to you as an alternative to becoming a full-time worker.

Labour costs are given as a major cause for the reduction in the size of the labour force followed closely in importance by the introduction of labour-saving buildings, machinery and methods. Next comes the changes in the way things are done on farms nowadays and, lastly, the decision so many workers on the land make to seek a higher paid or more attractive employment elsewhere. In this situation, the full-time worker has been the first to leave, often to escape isolation and to find social advantages many claim urban life can offer.

Finally, the Group of people who produced the report come down in favour of 'greater labour effectiveness' and how they think this goal can be achieved. You will not be surprised to learn *becoming a craftsman* comes first, followed by education for management and improved methods for recruiting and placing people on the land. The end result of putting their recommendations into practice, the Group believes, would be: *achieving a higher rate of increase in output.*

In bringing this chapter to a close I have been thinking about cash resources within the food growing industry to provide job opportunities as we move into more difficult times. Although there seems to be over-production of farm products in the EEC countries, Britain is still only 70 per cent self sufficient in dairy products. We could, if the national will were there, go all out to produce more of the butter, cheese and cream hard-working people thrive on. It also occurs to me that the two per cent levy that dairy farmers are paying with their milk cheques at the moment could be better employed providing job opportunities for young people. This levy was agreed upon at Brussels to discourage the over-production of milk. It is called a 'co-responsibility' levy and is a Statute of the European Economic Community. As a result it is binding on member countries, including ourselves.

Since it came into operation all dairy farmers have the 2 per cent taken out of their milk cheque by the Milk Marketing Board. The outcome will be another payment made to Brussels, an additional 22 million pounds sterling, to be precise. Cost factors like this discourage farmers from providing a job when the holding would benefit from the efforts of an extra worker.

8 Health and safety at work

Because so much has been said and written about the hazards and dangers of modern farming and horticulture a book like this would be incomplete if it did not offer some guidance. Agriculture, alas, has a bad safety record and all those who wish it well would not want to draw a smokescreen over the true situation. Neither would they try to suggest that because fatal and serious injuries and diseases affect only a comparative few, there is not the need for constant vigilance and care by every single person engaged in work on the land.

The habit of quoting figures, though very fashionable, can be counter-productive and often unlikely to achieve the result hoped for by the quoter. Certainly, as I have tried to reveal those things of importance to you as you consider a life on the land, statistics have not been thrown at you. However, in this chapter there is a need, right at the outset to tell you that between 1969 and 1973 over 35,000 non-fatal accidents in agriculture, on British farms, were notified to the Department of Health and Social Security. During that same period there were over 500 fatal accidents, many of them children. In addition men and women were very ill, and in some cases died, as a result of sickness and disease contracted working with animals, handling sprays and coming into contact with dusty and musty crops.

To the credit of the industry it has never been secretive about the fact that it comes very high up in the league tables of dangerous occupations. Very few of the annual agricultural shows, so much part and parcel of our farming scene, fail to offer a stand or exhibit aimed at keeping us all on the side of common-sense and taking reasonable precautions. Additionally, those groups and organisations involved with safety sit together round the table to devise ways and means to prevent accidents. Health and Safety Committees exist in most counties and have representatives of the work force, employers, and other interested parties to provide this essential umbrella sanction of the law. This is in the form of the Agricultural Division of HM Health and Safety at Work Inspectorate.

Any attempt to tell you what you must do to go through life healthily and safely once you start work on a farm, forest or nursery should be as near foolproof as possible. The attempt I am making takes the form of positive advice which, I hope, will prevent you from falling into lurking pitfalls and long-established snares for the unwary. There is no special order of importance.

Dust and fumes

There is some doubt about just how harmful most dust and fume you are likely to encounter may be. Certainly, the effect they have on individual men and women depends on a person's ability to resist their entry into the throat and lungs. Quite simply, these organs of our body have an ability to keep out much of the muck in the air we are inhaling but when this threshold of resistance is exceeded our health can be seriously affected—particularly if our breathing apparatus is already burdened with the soot and tar inhaled by cigarette smoking.

The golden rule is to wear the proper mask when levels of dust become apparent. As for exhaust fumes, avoid close proximity to them as much as possible. When fumes are present in discharged smoke, whether from an exhaust or a bonfire, check that everything possible is being done to keep the smoke emissions to a low level. Of course, there are regulations about these things and your employer has a responsibility in the matter. Nevertheless, it is *your health* which is at stake so always be on guard.

Sprays—herbicides, fungicides and pesticides

It has been said of weeds and their control that a good sharp hoe was the best herbicide; that is, alas, no longer true. Wherever you find yourself within the structure of modern food-production, spraying will be the method used to control weeds. It is convenient, easy and very effective but is also potentially lethal so never treat any spray casually. The role of chemicals in modern farming and horticulture is a major one and you may be called upon to apply them. Be absolutely certain that you read the makers' instructions on the package or container.

Fungus, an enemy of seeds, can attack growing crops, especially when weather conditions are favourable, and when crops are stored it can also attack. The Oxford Dictionary description includes a reference to a 'spongy morbid growth of excrescence'—sounds horrible doesn't it? It is, and so are the chemicals used to control it, unless they are treated with the greatest respect and care—follow the makers' instructions to the letter and, remember, new ones come on the market all the time!

Plant pests seem to be an army of amazing variety, of immense numbers, full of guile and cunning, able to move noiselessly and unobserved and all united in one grand purpose—to deprive mankind of his daily food. Like weeds and fungi, plant pests have to be considered within the delicate framework of what we call, the 'balance of nature'. However, as farming and horticulture are structured in the eighties you will find pests have to be attacked when they show themselves.

Slugs and snails immediately come to mind when we think of pests and certainly they move noiselessly and are not easily observed. The less notorious warriors in this army include aphids, eelworms, beetles, flies, moths, mites and spiders in all varieties, shapes and sizes. Plant pathologists, entomologists and nematologists are the names give to the scientists seeking to resolve the problems they cause. The more we intensify production and seek higher crop returns so do they become busier coping with increases in the pest varieties and sheer numbers.

Until the answer of natural methods is more developed, industry will continue to depend on what I can only describe as 'fire power', more potent and deadly pesticide sprays, to keep the plant pest army at bay. One such has the cryptic name of 245-T and many think it is capable of causing deformity, liver, skin diseases and cancer. Certainly, some Local Authorities have banned its use and, even though there is doubt about its lethal possibilities, there seems to be a strong case for not using it for spraying.

What remains to be said about spraying on the farm or nursery as far as you are concerned can be summed up as follows:

Always wear a complete set of protective clothing when using sprays.

Never expose any part of your body.

Regard undiluted sprays of any kind as deadly poison unless you have written evidence to the contrary.

Keep supplies of all sprays in a clearly labelled shed, outhouse or building and ensure it is not used for any other purpose.

Wash thoroughly after handling sprays in any way.

Never consume food or drink during spraying and not afterwards until you have stripped off your protective clothing and washed thoroughly.

Last, but not least, remember, it is always *your* responsibility to safeguard *your* health.

Tractors and machinery

So far in this chapter the emphasis has been on health which means, your health at work. Safety is no less important because carelessness when working on a tractor or with farm machinery and implements can result in maiming and death. In my job as a

Union organiser of agricultural workers, hardly a month went by without hearing of an accident at work and dealing with the inevitable consequences of such accidents. Always remember the most minor cut or scratch can lead to tetanus—I have seen what this does to people, too and it's not funny, believe me.

Also, remember a modern tractor is at least 40 times as powerful as a horse. In case someone disputes that figure, just remember how strong a horse is even moving at its natural pace. Tractors and machinery replaced men and horses on the land because of their ability to move much more rapidly than either.

Most farm accidents occur on tractors, hence the compulsory fitting of the cab or roll bar to protect the driver in the event of the tractor capsizing. Although it would be putting the cart before the horse to explain the finer points of tractor driving at this stage it seems right to advise any intending farm worker of the certainty of becoming involved with a tractor in day-to-day operations with animals, as well as the obvious role in cultivation. Therefore, when it is taken out for field operations it must never be thought of some kind of passenger-less car, far less a four wheeled motor-cycle!

As well as a tractor, trailer and tackle and equipment linked to both the back and the front of the tractor, you will come across many tools and mechanical aids that need to be approached warily. Power driven saws, whether of the bench type or portable, hand variety, have many lopped-off fingers and worse calamity to their credit—but only because the operative has not obeyed the instructions, allowing his attention to wander while handling them or, as is often the case with accidents, deliberately taken a short cut or flouted the rules!

Whenever wheeled tackle is being used great care is needed when approaching ditches or field headlands and hedges. Always see the rear as well as the front will get by. Never hesitate to throttle down the tractor and/or change into a lower gear if the situation calls for extra manoeuvrability. Steep slopes and gradients have claimed enough lives to fill the places in all our county cricket teams over the years but only because operating care was not exercised with the throttle and gear box of the tractor or the brake of the loaded trailer. Even crawler tractors with their wide, low slung bodies can turn over or slip on an over-steep gradient, so when you are eventually entrusted with one of these, the same rules apply.

There is plenty of fun and challenge to be had working on field cultivations and hauling jobs without indulging in dare-devil feats. Moreover, with all tackle now a major item of expense to the farmer and grower, he will not take kindly to the news that a breakdown or damage has been caused through taking what you claim was some sort of short cut, but which was really doing something reckless or stupid and putting your life at risk. It comes back to what has been said earlier in this chapter when dealing with sprays—'it is always *your* responsibility to safeguard *your health!* When it comes to handling tractors and machinery, however, the phrase should end with a change in the wording—'you life and limb' being more appropriate.

Mechanical aids

These two words give us a picture of manual labour being abolished. On the land this is certainly not the case although they have done much to lighten physical labour. Mechanical aids have not done away with one aspect of the working day since time immemorial—I refer to Lifting and Carrying. Bales of straw and hay have to be manhandled on occasions as do the draw-bars of trailers when they need hitching. On many farms and holdings fodder has to be carried to cattle housed in yards and buildings. These are tasks which can cause a rupture, strains or back injuries, any of which could lay you up for days, weeks or months. Even if you are possessed of great strength, avoid lifting or carrying excessive weights and always use the right method for lifting.

A lifting job should be sized up and not approached like a bull at a gate. Get a mate to help you if the load is excessive. When you do tackle any tasks make sure you know the right way to go about the job in hand.

Carrying weights can cause strains and be even more risky if you happen to fall with a load. Whatever the load is, get it high up on your back with enough of it on your shoulders to enable their width to give balance and stability. Here, as in lifting, be sure to know the right way to go about the job. With sacks of grain and seed proper lifting gear should be used.

In your training you would learn more and more about health and safety at work as the weeks go by. What has been written in this chapter will enable you to weigh up a little of what is involved before you take the plunge. In addition, I have had in my mind those many thousands who go on the land each year uncommitted as far as job training is concerned. Hence the final section of the chapter deals with the enforcement by law of health and safety standards.

Legislation

I have already referred to the Health and Safety Inspectorate dealing with the needs of agriculture. Some measure of the effectiveness of this organisation is contained in the annual report published in the middle of August 1980. A decline in the total of deaths and accidents in farming was recorded. Under the legislation now operating places of work have their own Health and Safety programme and this applies to farms and nurseries. When the number of employees justifies making the appointment, a worker's health and safety representative sees that the programme is kept to, as well as being on display. There is a mutual responsibility to see the farm is both a healthy and safe place for all.

Take safety seriously. Encourage those you work with to do the same. Never take a short cut by *not* putting on a guard and see to it that when the harvest hymn, 'all is safely gathered in' is sung, it really has been!

9 Employer and employee relations – the role of a Trade Union

Why a Trade Union? Because there is a close working relationship between those who work on the land it would seem at a first glance to be of little importance whether a worker belongs to a trade union or not. However, this view will not stand up to a very close examination once you have considered the reality of the working relationship between yourself and your prospective employer.

Farming and horticulture would be in a sorry state indeed if a powerful union did not exist to serve the interests of those who grow the nation's food. If this strikes you as being a strange statement remember that those who produce our food are in a minority within society as a whole. On the other hand, those who consume our food, being the majority, want to purchase this commodity as cheaply as possible. Another strong reason for the farmers and growers having a union is the long tradition in this country for food to be a cheap item in relation to the cost of living as a whole.

The National Farmers' Union (NFU)

To wean the community away from this attitude and secure a just standard of life for farmers is a colossal job and to ensure that the job is done properly the *National Farmers' Union* is well organised throughout the country, with about 90 per cent of employers, both farmers and horticulturists, in membership. Deservedly, it has the reputation of being the most powerful 'lobby' in the country. A lobby being the word used to describe group activity which affects the law-making process so that the interests of the group are served, and in recent years the NFU has not been above taking militant action on behalf of its members.

The National Union of Agricultural and Allied Workers (NUAAW)

As the agricultural work force is larger than full and part-time farmers put together there is a good case to be made for them to have an independent union, too. In fact, the Union has for many years served the interests of the work force. Those in farming and horticulture, forestry and a wide variety of rural occupations, from gardening to game-keeping; mushroom growing to Milk-Marketing Board employment; the whole alphabet of country-based occupations. Its main role is negotiating with the employer's representatives the basic agricultural rates (see chapter 10). The Union has a responsibility to provide worker's representatives on the Agricultural Training Board and its regional committees and County Wages Committees. With the passing of legislation aimed at protecting the interests of the work force in the country as a whole, it also represents members at Tribunals.

The NUAAW in the eighties has an important role to play in serving the rural worker and his family. Organised on an area or county basis, the members can be active within their local branches and district committees. These, in turn, initiate policies and activities which come before the area and county committees for sending to the Union's Executive Committee. Presided over by an elected chairman and serviced by a general secretary, head office staff and field organising staff, the Union seeks to assert a national presence to reflect the policies the membership requires.

A monthly journal, *Land Worker,* enables the members to express views both by letter and article. A vigilant editorial staff sifts out news items affecting the lives and well-being of the work force. The spreading of such news is assisted by the distribution of bulletins from the head office of the Union to area, county and district committees as well as direct to branches. It can be clearly seen that no rural worker needs to be cut off from the thinking of his work-mates. Equally apparent is the means at your disposal to exert influence to remedy any faults affecting the work force you may become aware of as your career progresses.

What of the future? Certainly the rural worker needs an organisation to protect his or her interests; to exert influence to secure fair working conditions and rewards, to provide representation on tribunals, committees and boards dealing with rural matters, and above all else, to strengthen the rural lobby and the rural way of life.

10 Wages and conditions

When you start work on the land you will begin to enjoy the wages and conditions that have come about directly as a result of years of struggle and subsequent law-making. What goes on now, ensures that you, and all your fellow rural-workers, enjoy conditions of employment likely to satisfy you, and more importantly, provide a career structure for the days that lie ahead.

The Agricultural Wages Act, 1948 provides for a yearly meeting of farmers' and workers' representatives to meet with independent members and fix minimum wages and provisions for holidays and for benefits or advantages. An Act of Parliament passed in 1978 could, to a greater extent, have more effect on your working life. It has the rather off-putting title: Employment Protection (Consolidation) Act 1978 and replaces an earlier Employment Protection Act 1975.

The earlier act spelt out a number of provisions to protect workers throughout their employed life. The latest one attempts to lay down more precise terms about exactly how this protection should operate. Ironically, it deals mainly with steps that can be taken when your employment has ceased. Beginning with remedies for unfair dismissal it then goes on to look at an employer's rights following suspension from work on medical grounds.

We never know in a working life when it might prove necessary to be aware of our rights. More importantly, though, we should always be aware of our obligations and what to give willingly and loyally in our job of work. The general view of country people, and one I share, is that legislation cannot replace our conscience in this matter. However, the laws now exist and we have to live with them.

Remedies for unfair dismissal After continuous employment for 26 weeks you can, if your employer sacks you, complain to an industrial tribunal. It is necessary to complete an application to do so, obtainable from the Department of Employment, forward this to the local office of industrial tribunals who will ask you to appear before them. They will decide whether, in fact your dismissal was wrongful or justified.

Itemised pay statements When you start work one important piece of paper which is of great interest to you is your pay slip.

Every week, or possibly month, when you climbed the ladder of success in farming, your employer is required to enclose one of these statements with the wages or salary you receive. Your pay statement will declare your gross earnings, including any bonus or overtime; then deductions for National Insurance and Income Tax will be shown and the amount payable will be the final entry. This, of course, should correspond with the actual cash you receive. Any other deductions, such as for a pension scheme, agreed between you and your employer, must be detailed.

Rights arising in course of employment detail firstly the payments which are guaranteed to you in the job you obtain. Your rights in respect of being unable to work on medical grounds are spelt out as is the matter of your membership of a trade union, including holding office in your union and devoting time to its activities as well as public duties that membership of a trade union might involve you in later on in your working life.

Maternity This section of the Act applies to females and covers worker's rights in connection with pregnancy and confinement. Maternity pay, and calculation of same, are dealt with as well as the steps that can be taken should difficulties arise at the place of work. These include complaint and appeal to industrial tribunal and the right to return to work after confinement.

Termination of employment When this occurs both employer and employee have obligations in respect of giving notice to correspond with the period you are paid for in the course of your employment. Your employer, on the other hand, is required to give you notice, or wages in lieu of, related to the period of time you have spent in his employment. A written statement should be given to you in the event of your being dismissed from your job.

Contracts of Employment

This is an important document spelling out all that has been agreed between you and your employer when your job starts. It should be in your possession within six months of the commencement of your employment. Your contract will detail the main provisions of the Employment Protection Act to supplement what has been agreed about your wages and conditions.

Redundancy payments Another of the provisions of the Act which, hopefully, will not come your way, though at the time of writing they are much in the news and something you must be wondering about in connection with your future working life. Should an employer find he is unable to continue to keep you at work you become entitled to a redundancy payment, if you are over eighteen and provided your period of continuous service is more than two years.

The payment is based on a calculation of one week's pay for every completed year of service up to the age of forty-one. Thereafter, until pensionable age is reached, one and a half week's pay is due for every year of completed service.

Insolvency of employers In the event of a job folding up because an employer becomes bankrupt any unpaid wages are treated as preferential debts. This is a difficult problem for any worker and, in fact, some 30,000 to 40,000 are affected annually. The number is growing at present. Claims for 4 million pound were outstanding when I was reading a recent summary of case histories affected by the insolvency, another word for bankruptcy of employers.

Should you find yourself in the unhappy position of having wages or holiday money owing to you this provision spells out what action has to be taken to recover the sum outstanding, up to a maximum of £800.

There still remain 33 items but, at the risk of being thought dilatory I am going to leave you to discover them yourself—having assured myself that they are not likely to affect you in your first years of work on the land. In any case, the membership of your Union will enable you to obtain assistance in the unhappy event of any of the situations referred to in the Act coming your way.

Agricultural Wages Act 1948

Under this legislation, new minimum wages came into operation on 21 January 1981. The Order, gives details of these, and provisions for holiday and for what are termed 'Benefits or Advantages'. These have wide-ranging significance throughout the rural scene, being the yardstick for determining wages to operate for the year. Most country-based occupations find the order has provided the base for fixing wages to be paid. Because of inflation, rates have gone up by varying percentages ever since 1948. In the last five years the rate of increase, like that of inflation, has been more marked.

To quote actual wage figures would be misleading as these become so quickly out of date. However, the latest table of wages can be obtained direct from the Agricultural Wages Board.

To avoid scratching of heads over the descriptions used in these tables I will explain these as simply as possible.

Appointment Grade I Applies to a worker having general responsibility for the management of a farm, a flock of sheep or a herd of cattle, the arable or horticultural operations on a farm or equivalent responsibilities. It is a condition of this grade that the manager has at least two other whole-time workers under his control.

Appointment Grade II A worker within this grade has day-to-day responsibility for the supervision of a farm or a flock or herd or the arable or horticultural operations on a farm or any other agricultural enterprise or operations of similar scope. Normally at least one other whole-time worker would be under his control but it is recognised that in special cases, owing to the extensive use of labour saving machinery or methods, there may be workers entitled to be in this grade even without a worker under him (or her, of course).

Craftsman This title has already been explained earlier and now that we are dealing with the financial rewards of working on the land you will appreciate why I stressed the need to train to obtain the qualification. The Craftsman is described as a qualified former apprentice who holds a certificate of completion of apprenticeship (in Scotland) or who holds a proficiency certificate.

As well as qualified former apprentices, workers who have obtained a craft certificate from their County Agricultural Wages Committee receive the premium payment on their wages. These can still be obtained by obtaining a declaration of competence from an employer but for the new recruit, either successful completion of ATB apprenticeship, or the obtaining of National Proficiency Test Council test certificates are a better way of qualifying.

Part-time workers play a vital role within the farming scene but it must be emphasised that there are two distinct categories at present. Those working over 30 hours per week regularly have a considerable edge, cash-wise, over those who enter into a working contract requiring them to do 30 hours per week or less.

Overtime rate The way farming is carried on in this country, or, for that matter, in most of Western Europe, means that overtime needs to be worked at times, known to all within the industry as coinciding with seed-time and harvest, milking and with stock-raising. When you are making arrangements with a possible employer about your working conditions be sure to discuss this fully and frankly with him. Equally important, is to discuss:

Provisions for paid holidays A whole-time and a regular part-time worker now has a minimum holiday entitlement of nineteen working days when he has completed fifty-two weeks employment on one farm or holding. This applies to working a five day week. Should you be required to work regularly on a Sunday additional holiday is given.

Because of the way holiday entitlement has become somewhat complicated you would be well advised to follow the advice given in the previous paragraph on overtime, discuss it fully with your intended employer in the knowledge that it is spelt out, to cover every possible situation, in the text of the Order. Trying to anticipate the questions forming in your own mind at this stage, I will mention here that sickness or injury do not affect your holiday entitlement as long as you provide a medical certificate.

Actual payment is based on the minimum weekly rates so the fact that you may be employed on a rate of pay that exceeds the minimum does not entitle you to this rate when you go on holiday. Here again, the best way to deal with that situation is when your contract of employment is being drawn up.

Provisions for sick pay Throughout the industry a uniform pattern exists for dealing with the problem of enforced absence from work through illness. In order to benefit from these provisions it is necessary to have been in continuous employment for 52 weeks or more. When this qualifying period has been completed you are then entitled to a maximum of 13 weeks sick pay, here again, at the statutory minimum rate.

The important thing is always to go to the doctor when you become ill. He will issue you with a certificate to send to your employer within four days.

Benefits or advantages are the last feature of the Order to be included in this brief summary. When a worker occupies a house, part of a house or a cottage belonging to his employer, a rent of up to £1.50 per week can be charged. Although such accommodation 'goes with the job', security of tenure is guaranteed under the Rent (Agriculture) Act, should employment end. However, it needs to be emphasised should occupancy of the house or cottage continue after this a 'Fair Rent' will have to be paid. Moreoever, if the employer can prove the property is needed for the efficient running of the holding, it has to be vacated as soon as 'suitable, alternative accommodation' is offered. Two years employment must be worked to qualify for this right.

Whole milk, if available, can be purchased by the worker at 3p a pint. It has been my experience that most herdsmen receive milk free but not as of by right: Potatoes can be purchased at the prevailing wholesale price of the district and where a single-worker requires accommodation and the employer is able to provide it, there is a recommended scale of charges.

It is however, generally accepted that very few employers are now able to offer accommodation on the farm.

The handicapped worker

Because the Agricultural Wages Order has a general note dealing with 'Permits of Exemption' this would seem the right place to deal with the needs of young people having physical injury or any physical or mental infirmity who feel, despite their handicap, they would like to have a shot at working on the land.

The major hurdle is, of course, finding a sympathetic employer. Such do exist and if any handicapped young person feels strongly enough about tackling the challenge every day at work will present to them, they will certainly make short work of finding such a sympathetic employer. It has to be said that the bulk of jobs in the daily round of work on a farm call for a 100 per cent fitness; the same applies to forestry. It is in horticulture that working opportunities may be found and if such arises the employer has the right to apply to the County Agricultural Wages Committee for a permit of exemption from paying the statutory minimum wage.

On receipt of an application it is normal for a sub-committee, consisting of workers' and employers' representatives to visit the place of work, meet with the two parties concerned, and if necessary, the parents of the worker. In the light of all the factors relating to the effect of disability on the performance of the job, they make a recommendation to the full committee on the amount of wage that should be paid. This is a percentage of the minimum wage, which if agreed is notified to both the employer and the permit worker. Normally the decision of the Wages Committee is accepted by both parties.

As an additional safeguard the permit is reviewed annually and, if necessary, a further visit is made to ensure that both parties are happy with the situation. The review also ensures wages are adjusted in the light of the annual review of pay.

11 Job opportunities in agriculture outside the United Kingdom

The lure of foreign travel affects young people more and more. Economy class air fares can take you to do a job of work as easily as they can whisk you away to that holiday of a lifetime. However, there is a major difference between working on a farm overseas and spending a holidy abroad. Most countries will welcome you with open arms for the latter but when it comes to the former the barriers appear.

A few decades ago a would-be farmworker could make his way to the Dominions of the British Commonwealth and be reasonably sure of finding harvesting work of some sort, joining a motley crew to apple pick in New Zealand, Tasmania or British Columbia or helping with sheep or beef cattle in Calgary or New South Wales. The fertile areas where intensive food production is carried on still need a willing and loyal work-force, though the emphasis is very much on experience.

If you have completed your school career and feel inclined to seek an opportunity abroad there is one absolute 'must' to be observed as a first step—do not attempt to go it alone, even though the nearby continent of Europe is so accessible and the member countries of the European Common Market place no restraints on working within their borders.

The Commonwealth

Every Commonwealth country has an office in London, as do sovereign states with whom Britain has diplomatic ties. Before getting carried away by the attractions of any particular country make contact with the Agricultural Adviser attached to their Embassy or Consulate. The laws relating to immigration are, in the main, restrictive but they change and modify and you could find a hole big enough to slip through—depending how strongly you feel about finding a foothold.

As a general rule you would be well-advised to consider finding a gap in the wall of legislation in a country where you have a close relation or family friend. At the risk of sounding hard I would qualify this and say, this should be an absolute rule, once you have made contact with the London office of the country of your choice.

The industry's newspaper carries advertisements of farm jobs in commonwealth countries. Usually, they stipulate experience or skill, but not always, so it is worth your while to get *Farmers Weekly,* the newspaper concerned. This seems the right moment to again mention that if you are interested in the industry, obtaining *Farmers Weekly* is a 'must', not merely for its situations vacant columns, but also for the articles, editorial and letter columns. These, together, give an overall picture of what is happening in farming and what goes on in the heads, and, sometimes, even the hearts, of those within the industry.

The Third World

We think of that part of the world where hunger or even starvation is always present as the Third World. The human needs of the ordinary men and women in these areas are often obscured by the highly technical approach of their would-be helpers. If you are one of an increasing number of young people who wants to do something on a practical level about starvation in Asia or Africa, by helping them to grow more food, be sure to weigh-up all the pros and cons before you attempt to take the plunge.

In the first place there are certain vital elements affecting the farming and horticulture in the Third World. Often the soil is deficient in essential nutrients on which crops depend. Secondly, rainfall is sparse and uncertain or else it pours from the heavens with such intensity that much is lost through run-off and encroachment, which really means the sand overwhelms the living soil, thus rendering it sterile.

If you are really concerned about the situation call in at your local Oxfam shop or write to them at Oxford or to some similar organisation such as War on Want, African Bureau or Christian Aid. In fact there are a host of groups concerned with the food crisis in the Third World.

Some awful mistakes were made in the past, dating from the notorious Groundnut mess-up in the late 1940s. It is now realised that the most effective way of helping is by improving on existing farming methods, which have changed little over the passing centuries. Limited mechanisation can enable ground to be cultivated more effectively and increase crop yields. Water for irrigation can be extracted by efficient pumps, provided the delicate balance of the water table is not put in peril. The harvesting and storing of basic foods can be assisted and large losses, common in the past, as a result of the inroads of vermin and insects, can be cut dramatically.

I hope that what is becoming crystal clear to you is that in looking for a job opportunity in the Third World you must become aware of what the real situation is as well as the needs. If, after you have come to grips with the problems that cause almost half the world's population to exist on a near starvation level, you decide this is where you want to work, do your year of practical training and then select a full-time course from the list of colleges after you have decided what to do and, *more importantly,* have a work opportunity in the country of your choice.

Finally, remember there are great needs related to the soil structure of undeveloped countries that you might be able to help with once qualified. I know of young people who have become involved in re-afforestation in Africa and been very welcome there, but what should be understood from the outset is that skill is a *'must'.* This is underlined in a statement published by Voluntary Service Overseas who offer opportunities to develop your own and a Third World country's experience. They say, I quote, 'You will probably be a qualified general agriculturalist/agricultural mechanic with some years practical experience'. You can find their address under 'Useful Addresses' at the end of the book and do remember that the pay is not high for the two year postings offered but air fares, accommodation and National Insurance/superannuation contributions are provided.

12 Your own farm or small-holding

It is not easy to recall the thought and longings of young people considering the giant leap to be taken when school or college life ends. The careers adviser may help to pin-point what we can do for a job but for many there will already be that strong urge not merely to work but to work for oneself. Certainly, life on the land seems to attract many whose dreams and longings go along these lines. 'Surely', they ask, 'it must be possible for me to farm on my own or make a living from a few acres?' There is wisdom in the old saying you cannot put the cart before the horse. Certainly, it needs to be said over and over again that, before anything else, to run your own place successfully you must gain experience on every level and develop qualities that are not required by those content to remain employees throughout their working lives.

This chapter will attempt to cater for those young people who can only consider working on the land if there is some likelihood that, having qualified, they can obtain a holding or farm of their own. Fortunately, it is a fact, men and women are acquiring places of their own *somewhere* in the country all the time.

What I can say to encourage you is I have known people who have done it; that is they have worked to become proficient and then set about acquiring their own place. Having reached this goal, by hard work and application, they have succeeded in maintaining their farming foothold or even extending it into a larger unit.

Those qualified to advise you would be very cautious and state, baldly, that the likelihood of getting a place of your own is slight indeed. If asked to comment on my statement that men and women are acquiring places of their own somewhere in the country they would admit to the truth of the statement and add that this was true—if you can raise the purchase price. What then are the present openings for those who want to obtain a place of their own for the production of food?

The allotment

Firstly, one of the humblest and yet, vitally important, sources of food production: is the allotment. This is really an extension of the garden and, like the garden, it provides an excellent opportunity to sound out your real interest in growing food. In fact, you can take over an allotment, as thousands of people in this country do, and cultivate it as you carry on your career on the land. The activity related to successful allotment holding is in contrast to what you are likely to be called upon to do during working hours; hand labour, working with small tools, as against large scale cultivation of land for crops or mechanised stock-rearing as in modern pig and beef units. Should your training years be spent in poultry or nursery work the intensive nature of your daily activity will be even more marked.

Being a parcel of land, the allotment, like the garden, has to be owned by someone, normally the local council, who are the people to approach regarding vacant plots. Varying in size, allotments are found on the outskirts of built-up areas, but in our cities they can be located in most unusual places. If you become interested and keep your eyes open, they appear in the spaces between railway track layouts, alongside reservoirs, in neat oblongs adjacent to sewage works, in fact almost anywhere unsuitable for buildings or commercial use. Hence, my advice to contact the local council for locations, availability and rental details.

Once having taken the plunge into tenancy you will find yourself alongside a group of enthusiasts willing to share their knowledge and experience with you. Your new neighbours will *not* share their tools and equipment with you. To acquire your essential tools of trade calls for a little working capital. What you expend in this way will prove to be a good investment for the future.

Firstly, you will need a small shed or water resistant locker to hold your tools and produce to avoid carrying your tackle to and from home in a wheel-barrow or vehicle; the former you are likely to need anyhow. Depending on the type of wheel-barrow you buy, the equipment to launch you on your food-growing career can be purchased for under £100. Minimal needs are a spade, fork, rake, hoe, marking-line, mattock, trowel and small fork. You might require a watering can, in the event of a hot dry, summer, and a net or similar protector if the local birds seem likely to be a threat to your newly-planted seeds. It will also be necessary to purchase seeds for the flowers and seeds you choose for growing.

This is enough detail to fire your imagination or confirm something of this nature is practical and within your grasp. It only remains for me to add that in the list of Useful Addresses at the end of the book you will find the national body looking after allotment holders, and to quote a recent statement in the House of Lords by Lord Davies of Leek: 'If people were not so toffee-nosed about keeping a pig or chickens on their allotments it would be a better Britain'. This was said during a debate on the future of allotments, when an assurance was given that any moves by local authorities to get rid of their allotments would have to receive the permission of the Minister of the Environment.

We must now leave small-scale husbandry for full-time food growing activity.

The Small-holding

Sharing with the humble allotment the feature of being owned by a local authority. The small-holding differs, however, in that in recent years it has been under much greater threat as local authorities have seen it as an asset capable of being put to more profitable use. As the County Councils sell them off to raise much needed revenue, small-holdings tend to be harder to come by from this source. Fortunately, other opportunities exist and chief among these are those of the *Land Settlement Association,* a co-operative venture owned by the Ministry of Agriculture, Fisheries and Food.

The LSA provide an ideal opportunity for would-be small-holders. Requiring from their tenants proof that they possess the skills to plan and produce for food growing, the applicant for a holding has also to convince the Association of his ability to cope with the stresses and tensions associated with running the enterprise.

On their part, the LSA offer comfortable, up-to-date housing on the holding, though, mostly, these are small properties. In addition to this advantage must be added the undertaking all small-holders of the Association are given about their produce. This is taken to local packhouses, usually nearby, thence to market for sale at the prevailing price thus lifting one of the big burdens of a small-holder. As well as grading and marketing the produce, a plan is prepared for individual holdings based on the common cropping plan of the estate where it is sited. This plan shows the working capital required, the first year income of the tenancy and confirms that most buying and all selling must be done through the Association.

Working capital for a small-holding is a very different matter to the acquisition of hand tools described in the allotment section of this chapter. The Land Settlement Association requires an initial working capital of between £16,000 and £20,000 depending on the production plan. However, 75 per cent of this can be borrowed and the Association will arrange for an agreed balance of the initial working capital to be borrowed from a bank or the Ministry of

Agriculture, to whom the Association is responsible.

Hopefully, by the time you have completed *at least* five years practical agricultural experience and are between 25 and 40 years old (the preferred age range) the present high, borrowing rate will have come down a little. Finally and before we move on to owning a farm, remember local authority holdings provide permanent buildings and certain other fixed equipment. Working capital is always required for the purchase of livestock, feeding stuffs, glasshouse, implements, etc. When the time comes to take the plunge application should be made to the County Land Agent at the offices of the Council in the area of your choice, or more probably, the area where a small-holding is available.

Owning a farm

Unless you are the child, or close relation, of a farmer or expect to marry into a farming family, to get your own farm calls for many character qualities, exceptional practical ability, a hard head for business and a willingness to devote your life to the acreage you are enabled to lay your hands upon. Of course, all these qualities are required if you are going to inherit a farm, but, unlike those who can inherit, you are going to need that rare commodity, working capital, in a much larger sum than for a small-holding. At the moment the banks seem keen to finance any farm development, except starting from scratch. However, when the opportunity arises, contact with the Agricultural Advisor or Representative of one of the 'Big Four' banks may prove to be to your advantage. Given a greatly improved situation in the country as a whole you are still likely to be required to raise a large proportion of your required working capital.

Back in 1945, author and farmer, R M Lockley, put it like this, 'Some capital will be needed, but above all the considerations of capital, experience and situation, application to work is the first and last requirement, for successful farming demands unremitting thought and labour. The farmer and his wife and children must make up their minds that their leisure and their work are the same thing, and find joy in the thought'.

Some capital! An average price of farmland is pretty meaningless as there are great variables and regional differences that affect the situation. When a farm becomes vacant it is normally offered for sale by agents of local standing, or in the case of large farms, by the land and estate agents who have a national reputation. In either case the properties are put up for auction or

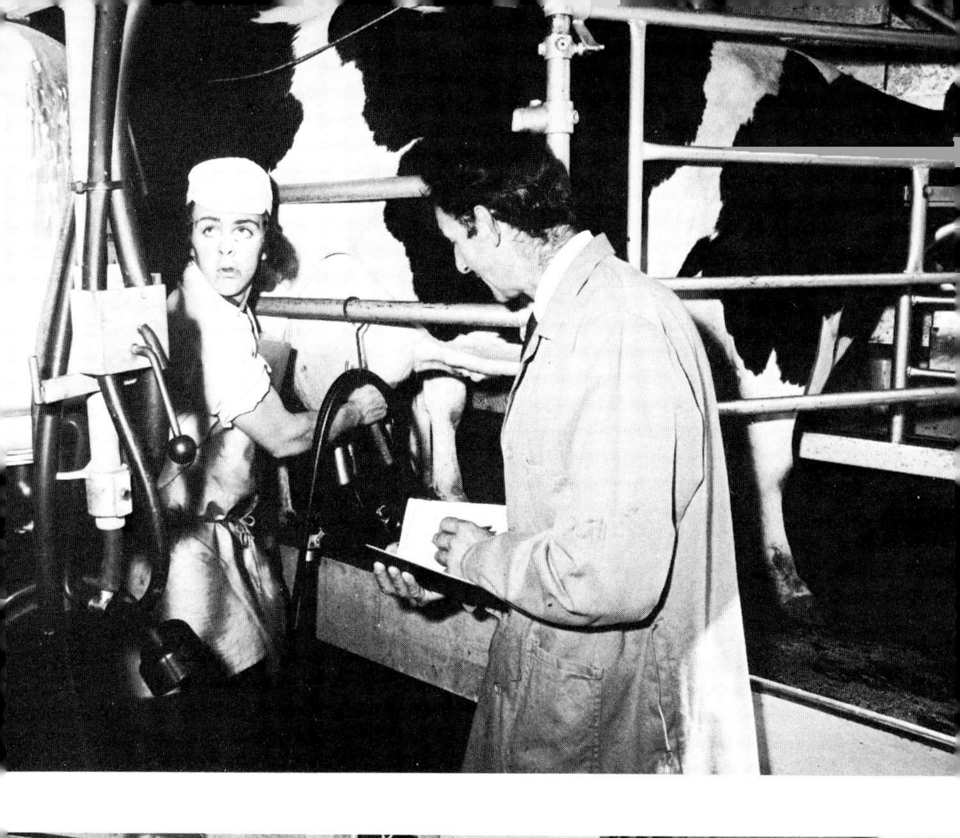

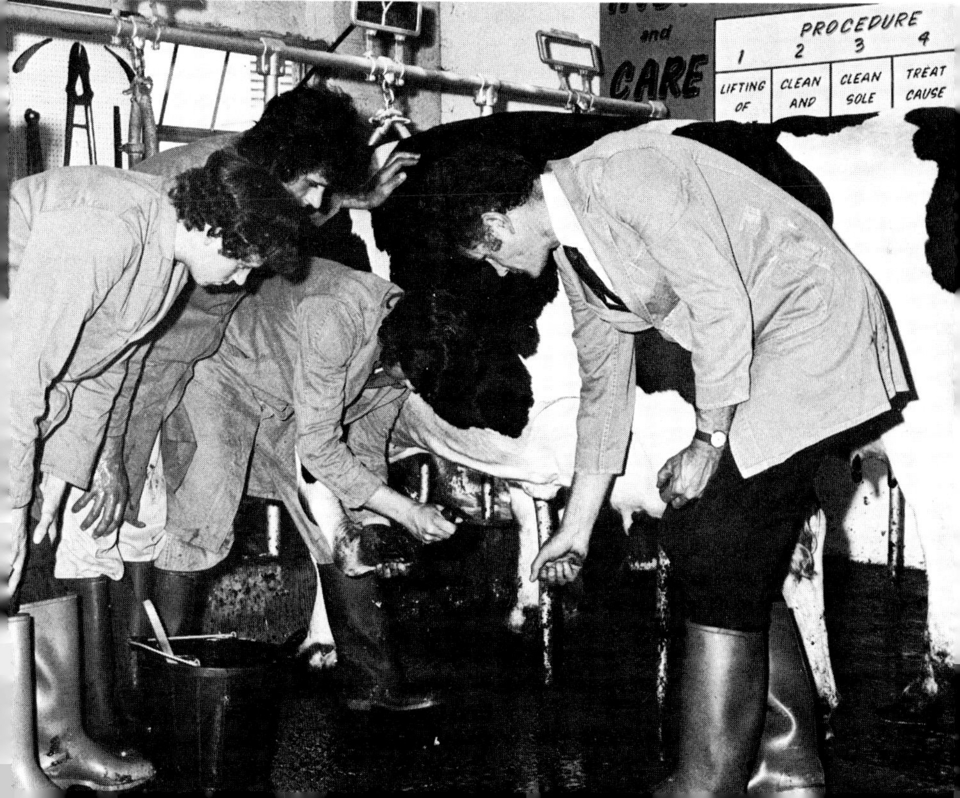
and
CARE
PROCEDURE
1
2
3
4
LIFTING OF
CLEAN AND
CLEAN SOLE
TREAT CAUSE

MTW

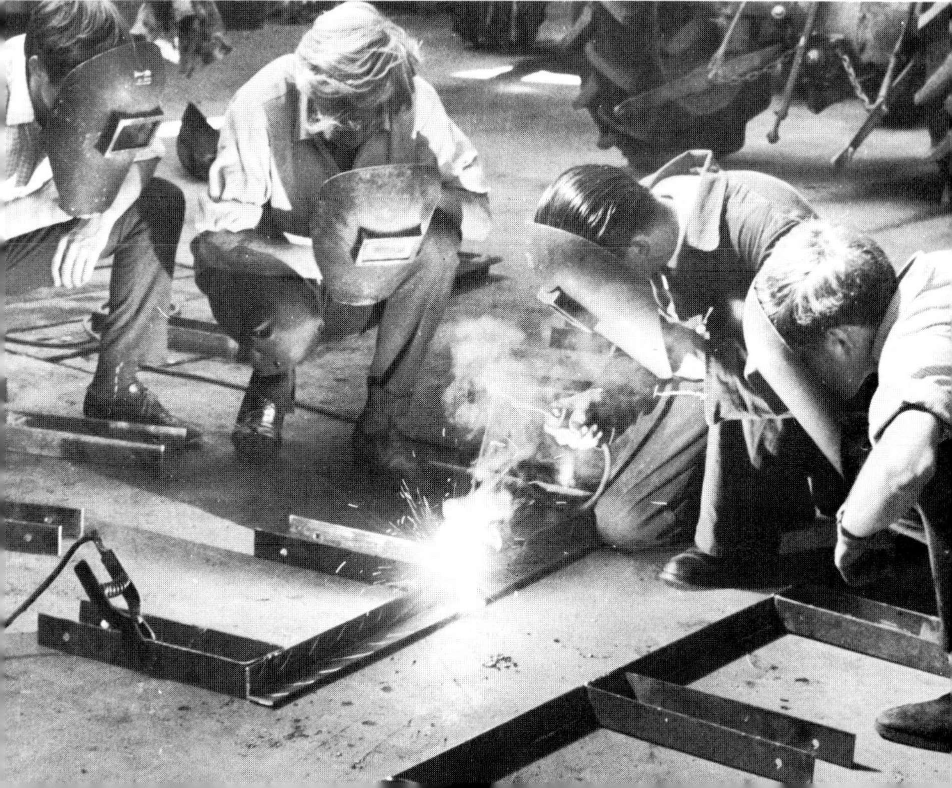

DEPOSIT 15P

offered for sale at a fixed price.

Farmers Weekly, devotes up to ten of its pages to advertisements of farms for sale by the said agents. However, only a few square centimetres of space are filled by details of local authorities and owners offering farms on a letting basis. A typical advertisement will run like this:

> EXETER 7 MILES (in large type because this is a favoured area and one where farms change hands frequently). On the edge of Haldon Forest in a superb highly accessible, totally secluded position. An outstanding Residential Holding small Stock Farm with exceptional modernised house (6 bed. 3 recep. etc). Excellent bldgs and 39 acres (16 hectares) £115,000. Further 48 acres (19.4 hectares) available.

That exact quote has only one addition from me, the bracketed reference to the location! With land at over £400 per hectare you can easily calculate that to move into such a place would require over £200,000 by the time you had stocked the place and obtained all the machinery and tackle required. The income to be derived from such an acreage would be about the same as a skilled worker's wage, assuming first class management. Of course, you could hope to be married to an understanding wife and take in visitors, a common practice in the West Country and one that permits small farms to survive.

Also in the *Farmers Weekly* are other examples of farms for sale in many parts of the country, some of over 500 hectares. One with a small bungalow and 50 hectares of grassland plus excellent farm buildings and modern milk parlour in Lancashire caught my attention. It was being offered for a quarter of a million pounds! All I am attempting to do by quoting these two examples is to bring home to the aspiring farmer the reality of the financing aspect. Having dealt briefly with the qualities and character required to become a farmer and having pin-pointed the simple facts of buying a property to carry on the craft of farming, let us now assume that you have faced these two preliminaries, convinced yourself that you can cope with them and now want to know more. Certainly, to have a little knowledge at this stage may help in confirming the certainty you are building up over coming to grips with being your own boss.

Making your personality fit the challenge Develop your sense of humour so that you can laugh at yourself as well as others. Leadership implies an understanding of people and the ability to get alongside them—shoulder to shoulder if they are working with you or for you. Don't assume you have the gift of management because you can organise; learn it thoroughly and in the company of others, preferably at an approved course of studies such as those offered by the Agricultural Training Board.

Never forget the variables of weather, livestock health, market prices, policy decision making, and remember the EEC have made some very complex decisions that affect farmers.

Being business-like does not consist entirely of having the business instinct. As a farmer you will have to understand fully the pitfalls and the very real advantages of various subsidies and grants. You should understand capital transfer tax, overdrafts, rates of exchange, after-tax profits, your income tax, cash flow, hire purchase and many other things unrelated to skill in producing crops.

Developing trends play an important part in farming. What is certain is they will play an increasingly important part in the years ahead. I have already made a brief reference to providing farmhouse accommodation but, also, more and more townspeople are beginning to seek greater understanding of what goes on in the countryside, hence grants are available in setting up facilities to help people obtain this understanding. Picnic areas, camping and caravan sites immediately spring to mind and the need to explain farming methods to visitors. Grants are available for these as well as for constructing farm trails, planting amenity tree belts and laying on what have come to be called 'Farm Open Days'.

What better way of ending this chapter than by reminding you that despite all the pessimism about our declining countryside, in 1978, 650,000 people were employed in agriculture. This is 2.7 per cent of our population in Britain and the smallest proportion in the EEC. Less than half of Germany, less than a third of France and less than a fifth of Italy. Surely, given the will to become more self-sufficient in food growing and less dependent on costly and precious fuel, the number will increase and more farms come into being to provide working opportunities. Develop your will to become a farmer or small-holder in your own right and help to create that climate of opinion so necessary if we are to re-populate the acres of 'England's green and pleasant land' once more.

13 The role of forestry

When, in the course of this book mention has been made of trees and woodland, one vital fact has been omitted. We have less of our land area devoted to forest and woodlands than any other European country, apart from the Republic of Ireland.

Fortunately, this is not the whole story. Amenity tree planting, the use of shrubs and trees in urban landscaping, and enthusiasm for tree belts and small woodland lots as reserves and game areas in the countryside, are all helping to reverse what had become a disturbing national trend. All the more serious because forest and woodland, spinnies and coppices, have played such a major role in the shaping of our history as a nation. One has only to think of the great 'wooden walls'—ships that made us the major maritime power in the world; the furniture of Sheraton and Chippendale; our half-timbered houses built up to the sixteenth century and many still standing to this day; our farm tools, wagons and carts. All these and a host of other vital, domestic products speak of the role played by our woodlands. But there is more to it than that.

Trees play a key role in the air we breathe, the climate we enjoy and in the maintenance of the delicate balance of nature, plant, insect and animal so essential to the living soil. However, the title of this chapter relates to the role of forestry in providing careers on the land. Its wider role is the subject of many volumes, written by men who love our native woodlands. A few of these titles are listed at the end of this chapter. Within their pages can be found the absorbing story of the forest floor and its myriad of inhabitants.

Forestry, according to the dictionary, is 'the science and art of managing forests'. It includes everything from planting and tending plantations of conifers in the north of England and Scotland, to forming small broadleaved (or deciduous) woods for amenity and game management on the lowlands of Britain. The Forestry Commission—a State enterprise—is the largest, single employer within the woodland scene and involved in the management of much of our forest acreage of which they own half. A leaflet issued by the Commission describes the raising of millions of seedlings and transplants in forest nurseries to the production of large standard trees for the streets and open spaces of our towns and cities; and the felling and transporting of roundwood to sawmills and to factories producing pulp or chipboard. It describes the bases of forestry, silviculture and arboriculture, which requires intimate knowledge of trees and their culture. The close links with agriculture are also stressed as are the needs to have a business-like approach to the production of wood. The career opportunities are then dealt with and, because the opportunity to find work in the forest is likely to be with the Commission, you should know about these opportunities.

The Forest Worker

The Forest Worker is engaged in the manual work and operating of machines. You would find yourself called upon to tackle a wide range of activities including fencing, planting, draining, weeding, pruning, timber harvesting and nursery work. You would be expected to acquire a high degree of skill with forest tools and machines. These would be similar to those used in farm work but with more emphasis on the chain-saw.

Training is given in the various skills required to enable you to qualify as a forest craftsman. Financially it is, of course, an advantage to become a craftsman. Having reached craft status there are occasional opportunities for promotion to ranger and foreman. A foreman who holds the City and Guilds Certificate in Forestry and the Certificate in Supervisory studies becomes eligible to study for the City and Guilds Forestry Stage 3. Holders of these certificates can compete for posts in the Forester Grade on level terms with candidates who hold the ordinary national diploma in Forestry.

Before telling you about the job of Forester it is worth mentioning that in connection with forest protection, wild life control and conservation and recreation, vacancies occur for rangers. Although such opportunities are few they are normally filled by forestry workers who show an aptitude for this kind of work. Very occasionally, vacancies arise for warders to manage one of the Commission's Camp sites in places where forestry activity is carried on in a holiday area. Availability of jobs varies from region to region and applications should always be made to the district office of the Commission where you live. Details can be found in your area telephone directory.

There are no opportunities for apprenticeship but on being accepted for a job as a forestry worker you would be given every help and encouragement to make progress along the lines outlined earlier. You would certainly find yourself alongside men in a gang of varying ages and levels of experience and skill—an excellent way to learn any job. Unlike farming and horticulture, living-in jobs are almost non-existent though in some regions board and lodgings away from home can be found.

Foresters

As technical managers Foresters are responsible for planning work programmes, supervising and training workers, estimating costs, protecting forests and for relations with neighbouring landowners, organisations and individuals who wish to use the woods for sport and recreation. Once experienced, they specialise in training, work study, wild life conservation and research. Training is given at Cumbria College of Agriculture and Forestry, Newton Rigg, Penrith, Cumbria and consists of a three year 'sandwich course' (ie one year practical between two years theoretical), leading to an Ordinary National Diploma in Forestry.

This course requires a minimum of four O levels, or four Grade One CSE passes, two of which should be science subjects and one which tests your command of English. You would need to contact the Principal of the Cumbria College for full details of this course and remember that at least two year's practical forestry experience is required before you can start. An offer of a place at the college will be subject to your gaining this experience and when evidence of having a provisional place at the college is in your possession the Forestry Commission will consider you for employment in a forest for your two years practical experience. In this case you application should be sent to: The Chief Education and Training Officer, Forestry Commission, 231 Corstorphine Road, Edinburgh, EH12 7AT.

Students grants may be awarded at the discretion of your local education authority and I must remind you again that the Commission does not provide apprenticeships in forestry. Neither does it sponsor applicants to Cumbria College.

The Forest Officer

This officer is responsible for the planning and control of planting and conservation work as well as the preparation of woodland produce in groupings of forests known as districts. Also, for private woodlands within their districts and possible involvement in the acquiring of further land for forestry. Some go on to specialised work in recreation, training, research and development. This is a job to bring you into contact with the general public as well as 'key' people. Being in a sense what is called nowadays a 'plum' job, service starts by assisting a senior forest officer for training and to gain experience. At the present time there are, according to the Commission, excellent prospects for promotion to Forest Officer Grade I after about five years and there are good promotion prospects.

Top managerial posts

Promotion to higher grades is dependent on merit and Forest Officers with a high level of ability, can look forward to being part of a team of senior managers responsible for general forest management, harvesting and marketing, estate work and administration. These duties could be undertaken in any of the regions of the Commission, which are controlled by Conservators.

Entry to the highest positions in the Forestry Commission service depends on success at an annual competitive interview, competition is severe as only a few recruits are required in any year. Candidates must have either:

(a) a degree or diploma in Forestry or closely related subject having a forestry content;
(b) a post graduate degree in a Forestry subject;
(c) an equivalent or higher qualification.

Courses to reach degree standard are provided by the Universities of Aberdeen, Edinburgh and Oxford and at the University College of North Wales at Bangor to whom application should be made via the Registrars. Full information including the syllabus of the course would also be available from this source. Ability to drive a car is essential.

Opportunities in the private sector of forestry

Of the 33,000 thousand jobs in forestry in Britain over 23,000 are to be found in the private sector. Although an approximate figure it will serve to illustrate the extent of job opportunity, though it does not point to the number of vacancies waiting to be filled at present.

W E Hiley CBE, MA, one of the 'giants' in modern forestry practice, looked forward to a future in which the State-owned woods, and all other woods in the private sector, would produce, in partnership, the pattern of a joint forestry economy which might, in time, offer a model for other nations to notice and perhaps to emulate. In turn President of the Royal Forestry Society of England and Wales and of the Society of Foresters of Great Britain, Hiley wrote and spoke about his forestry theories, then, determined to put them into practice, he took over tne management of about 800 hectares of mainly unproductive forest at Totnes, in Devon, now Dartington Woodlands Ltd.

From the time he made that decision in 1931 until his death, Hiley, worked to establish a model of what private forestry should be, a period of 30 years during which many sat at his feet learning the arts of silviculture and arboriculture. There have been other giants too who should be mentioned here: Firstly, *Herbert L. Edlin,* whose long and honourable career as a Forest Officer, included rubber planting in Malaya, work in the New Forest and 30 years as Publication officer for the Forestry Commission. His enthusiasm for all the wonders of trees, plants and the countryside, as well as the love of woodland crafts and country people he revealed in his books, won him a special place in the affections of many in both the private and state sectors. Some of his books are listed at the end of this chapter to encourage you to read them and because in a volume of this size it is impossible to deal fully with Forestry.

Secondly, *Richard St Barbe Baker,* founder of *The Men of Trees,* now a world-wide organisation seeking to give trees the place of importance necessary if our teeming planet is to survive. He has served as an assistant conservator of forests in Kenya and also in Nigeria in Colonial days. A tribute paid to him by the late Earl of Portsmouth sums up the importance of what he has done, 'He is not a professional writer, but an Earth Healer. If we do not heal the earth which we have afflicted and exploited so grievously, then there will be neither health nor peace among the nations'. However, he *was* a writer of considerable merit and for this reason I have listed his most important book.

Private contracting

Apart from private enterprise there are many sources of local employment within the private sector of forestry. Among them contractors up and down the country whose existence you would only be aware of if you lived where they operate, usually thinning, felling, extracting, etc, in Forestry Commission woodlands.

Contractors rarely include preparation for planting and tree planting in the scope of their activities. A more varied life is often to be found on the remaining larger estates of this country where forestry and woodland are maintained, planting taking place to fill the gaps in woodland areas that occur after felling and clearing operations. These estates vary in size and in the number of job

opportunities provided. For details write to one of the following:
Economic Forestry Group, Forestry House, Great Haseley, Oxford, OX9 7PG.
Timber Growers' Organisation Ltd, National Agriculture Centre, Kenilworth, Warwickshire, CV8 2LG.
Tilhill Forestry Group, Greenhills, Tilford, Farnham, Surrey and at Old Sauchie, Sauchieburn, Stirling, Scotland.
Forest Thinnings, 20 Turk Street, Alton, Hampshire.
The Scottish Forestry Training Group, Room 122, Forestry Commission, 231 Corstorphine Road, Edinburgh, EH12 7AT.

Wages

Wages are negotiated annually for Forestry Commission workers and are not binding in the private sector where they fluctuate considerably. However, it is generally accepted that in private sector employment, housing and other benefits are offered.

For details of the Forestry Wages structure write to the Forestry Commission.

Camp Wardens, Forest Clerks, Senior Rangers and *Rangers,* of whom there are few employed, receive rates relevant to the above. Piecework rates, operating when productive work is being carried out, enhance earnings and overtime is paid for hours in excess of 40 per week. These rates of pay apply to Foresters who begin their training by working in the woods alongside forestry workers. When qualified, their starting salary is around £5823 per annum, rising to some £7560.

Holidays As from 1 November 1980 the annual leave allowance for forest workers in their first ten years of service will be seventeen and a half days per year. Increasing by one and a half days after ten years of service. A holiday supplement to wages will also be paid.

Clothing Working in the forest poses some special requirements as far as clothing is concerned, with the need for protective headgear springing to mind straight away. The need to use the chain-saw for a variety of operations calls for special trousers, gloves, ear and eye protection and safety boots as well. You will begin to realise that in the woods, clothing is very closely related to health and safety at work. Hence, the Commission and employers in the private sector arrange issues of clothing with this in mind and with emphasis on protection from the weather. You will not need to concern yourself with kitting-out in the way you have to for farmwork; once at work your clothing needs will be met.

Forestry Safety Council

Because of special safety hazards, work in the woods requires the observance of detailed codes of safety practice. The Council has prepared simple, illustrated leaflets to guide the workforce in tackling any forestry operation. Working in the woods is no more dangerous than other rural jobs. However, because of the increasing use of mechanical aids and the passing of laws about health and safety, the wise precaution has been taken to introduce leaflet guides dealing with most forestry operations. Like protective clothing, these will come your way very soon after you become a forestry worker, and are an important 'aid' to training.

Having considered the wages and conditions I will add the warning you have already come across in this book. The material rewards, when examined coldly, will not woo you into our forests and woodlands. You will need to feel something of what Henry Thoreau experienced in the middle of the last century, called the 'genius of Concord' (a town in eastern U.S.A.). He was a gifted visionary who left the life of the town to live in a wooden hut of his own construction in the woods of Maine.

During the two years that he spent there, alone with nature, his life was one of great simplicity with the inhabitants of the forest as his close companions. He writes of 'cutting and hewing timber' and of growing his own food. During the long, dark and bitterly cold New England winters, he read widely as well as coping with daily chores; at all times marvelling at the wealth of sights and sounds in the forest. The book will give you a better idea of where the real wealth of living is to be found—and where better than in the woods? Finally, essential reading which you need to fill the gaps in this survey of forestry, starting, of course, with Thoreau's *Walden* which, like the books listed below, can be obtained from your local library or bookseller:

Richard St Barbe Baker, *I Planted Trees,* Lutterworth
W E Hiley, *A Forestry Venture,* Faber
H L Edlin, *England's Forests,* Faber
H L Edlin, *Collins Guide to Tree Plant and Cultivation,* Collins

To give you some idea of the wealth of crafts and skills linked with work in forestry, no book is more readable than: H L Edlin's *Woodland Crafts in Britain,* David and Charles.

14 A mixed bag of rural opportunity

There remains a number of job opportunities within the countryside that I propose to group together into this chapter. Not because they are any less important than farming, horticulture or forestry but because, in terms of numbers, only a few young people can expect to find their life's work within them. Some have already been mentioned and, because of its importance in the food chain, I am going to deal first with:

Beekeeping

My personal contact with bees has been limited to helping to collect a swarm, listening to beekeeping friends air their knowledge and collected wisdom grouped around the hives, and, most agreeable of all, enjoying the many varieties of honey available as a nutritious and very tasty food.

Much as I would like to be able to write of job opportunities with bees, enquiries I have made convince me that such are practically non-existent. Experts have told me that full-time work in commercial beekeeping has gone because of the nature of modern farming. Less clover pasture and the use of sprays seems to have caused this state of affairs.

However, the case I put for the ownership of an allotment, when a larger agricultural holding is ruled out, compels me to say something about it and advise your if you are one of that small number determined to establish your own apiary. Proceed no further unless you are patient, gentle and willing to endure some stinging. Add to this the certainty that you are really keen and you could be suitable material for joining the distinguished company of British Beekeepers.

Many more bees are needed in towns and suburbs, as well as in the countryside, to ensure adequate pollination of our crops of top fruit, soft fruit, seed plants and a variety of field crops, also to safeguard our wonderful heritage of wild flowers.

All this means, of course, becoming a beekeeper in your own right. You can keep up to maximum of 40 hives and be regarded as a domestic beekeeper; keeping up the tradition of a cottage industry that has flourished in Britain for centuries. Remember, until Tudor times, honey was our only sweetener, and even earlier, our principal beverage, mead, was derived from honey.

Every country has a beekeeper's association and an agricultural college or farm institute where courses on beekeeping are given. I would strongly advise anyone to establish those two contacts before proceeding (see Useful Addresses). A large capital outlay? No. 'You could be in business' for less than a £10 to cover the purchase of your first hive and swarm of bees, numbering at least 30,000 with a 'queen' in supreme authority.

Gamekeeping

This is as old and as important a calling as beekeeping, with the added advantage that jobs and careers are still available to those who can face a somewhat lonely working day, going home at night to a cottage normally outside the village and often isolated. Excellent eye-sight and hearing and a good level of physical fitness are required for this, all-weather and seasons, occupation. The 'game' to be kept, and reared with care and devotion cover much of our woodland life—pheasant, partridge, duck, hare, rabbit, deer—all these come the way of the keeper as he seeks to keep coppice and woodland well-stocked for those prepared to pay high price for a good 'bag'. Woodcock and grouse are rarer breeds and for this life you need to be knowledgeable about ,these birds and beasts as well as the creatures that prey on them. No writer is more knowledgeable that Ian Niall, who writes weekly in *Country Life* as well as producing a range of books that might confirm your interest in gamekeeping as a job for life.

You should read a down-to-earth booklet detailing the nuts and bolts of the job. Published by the Hampshire College of Agriculture, Sparsholt, Winchester, Hampshire, it is one of a number of useful leaflets that deal with all aspects of the work. Moreover, they come from England's foremost game county.
Write to: The Information Officer, *The Game Conservancy,* Fordingbridge, Hampshire, for help and advice about pupil and apprentice opportunities.

Water Bailiff

A similar occupation is that of Water Bailiff or Water Keeper, whose job it is to keep our rivers and streams stocked with game fish, principally trout and salmon. With so many threats to our streams from pollution as well as the more obvious poaching this too, like gamekeeping takes a long while to learn. There are less opportunities but initially you could get a *Career Outline* free from: *Career Occupational Information Centre,* The Pennine Centre, 202 Hawley Street, Sheffield, S1 3GA.

Having ventured into water-based occupations we can naturally move on to one offering more and more job opportunities nowadays.

Fish Farming

This is a form of husbandry resembling intensive stock-rearing. Like the chicken, the trout has ceased to have undertones of luxury living, and is in strong demand from hotels and restaurants as well as shops. In recent years fish farms have literally mushroomed and, certainly, they are offering more and more job opportunities.

Though by many to be a new development, they are, a modern version of the fish pond husbandry of the monks, introduced into this country before the Middle Ages. To avoid offence to those involved in the practice, I would hastily add that whereas the monks were concerned with the rearing of a regular supply of carp or trout for meatless Fridays, the modern fish farm is concerned with the production of up to 150,000 trout at a time.

Breeding, hatching and rearing ponds are constructed alongside a flowing river or stream, as free of pollution as possible. These ponds, being fed by the stream are, in a sense, an extension of it, but with the added advantage that the natural enemies of the fish can be, to a large extent, be kept out. Thus it is possible to rear them in their thousands rather than singly or in pairs. With public appetite for trout existing at the moment, plus the demands of the catering trade, it would appear certain this form of husbandry will develop and job opportunities increase.

The Pennine Centre, mentioned earlier in this chapter, publish a leaflet, 'Fish Farming No. 43' while the Hampshire College of Agriculture similarly mentioned offer courses and full information on this new form of job opportunity. Lastly, the National Farmers Union have created a *Fish Farmer's Specialist Branch* at their London headquarters, confirming the importance of the activity in the world of food production.

A Career with Horses

Where do you start a consideration of the horse and its role in the life of our nation? In this book I have been on the point of digressing whenever it has been necessary to mention this noblest of all our animals and certainly one for which I feel strong affection. It now remains to say something about the important role the horse still has in farming, forestry and horticulture; while at the same time omitting reference to the other activities and uses we associate with the animal. These, have volumes written about them and provide many job opportunites. I refer, of course, to the leisure pursuits the horse provides; racing, hunting, hacking, trekking, jumping and pulling a great variety of wheeled transport.

When cultivation of the land moved away from hand operations, the horse superseded the oxen in providing power. Only when oil became cheap and plentiful did the internal combustion engine take over farming operations. Until the outbreak of World War Two it was customary to see the horse ploughing, cultivating, sowing and reaping. As well as coping with field crop operations, it made forestry operations feasible; the horse-drawn pole wagon, with its team of six or more horses, being a quiet, clean way of extracting heavy felled trees.

The horse is as demanding food-wise as the tractor is for its fuel. It was the need for fields rich in clover to provide this fuel that so helped bee-keeping. The existence of the horse,and its major role in food growing, also ensured the need for a much larger rural work force than we have nowadays. In fact, both in France and Italy, where the horse has continued to be used in agriculture; over three times as many are employed in the former country as here and six times as many in the latter. More importantly, the skill of horse breaking and rearing has been maintained without interruption in these countries. Father hands down to son the skills of working the horse in its many varied tasks to an extent no longer known is this country.

With fuel oil increasing its price tenfold and becoming scarcer (despite the miracle of North Sea Oil) we have to give serious thought to using the horse again in a wide spectrum of farming operations because it is more economical to do so. Moreover, many more people will be needed to learn the skills of horsemanship again.

During the mid-twenties and thirties and forties of this century, when the tractor assumed its ascendancy the working horse was threatened with extinction. The fact the Shire or Great Horse of England, survives, is due to a few far-sighted individuals and breweries having common sense to realise that the horse still has a practical application even in the mechanical age. Similarly, our other great breeds were maintained—just! Hence the Suffolk, the

Clydesdale and the Percheron can still be seen working at ploughing matches and various agricultural shows. The National Shire Horse Show, held at the East of England Showground, Peterborough in March each year attracts hundreds of stallions, mares, geldings and foals of that truly magnificent breed to demonstrate the working horse is back in business.

Because the horse is more economical for so many farm jobs irrespective of the size of the farm, training courses are available to learn harnessing and handling, ploughing, feeding and management, common disorders of the horse and their treatment with the opportunity to receive follow-up training for about 150 students in the last five years, being attended by both employer and worker wishing to add the skills of horsemanship to their other abilities. Discussing this heartening trend with Mr. Charles Pinney of Cotswold Farm Park, Guiting Power, Cheltenham, where the training is given, I found students come from Devon, Wales, East Anglia, Sussex and Lancashire; both male and female. In fact, Mr Pinney stated that, if anything, women showed more aptitude!

The Agriculture Training Board sponsor these courses in much the same way as other Craft Skills schemes, catering, in the main, for those already involved in agriculture who have come to appreciate a full understanding of horsemanship will stand them in good stead in a world where supplies of oil are beginning to dry up and, in all farming operations, threaten to become prohibitive, cost-wise.

Be assured, career opportunities do exist and are likely to become more plentiful. Although not directly related to farming there are job opportunities with breweries, who maintain the splendid stables of draught horses already mentioned. More and more farmers are thinking of going into horses while many have never lost their certainty that there is a place on the land for the working horse.

Lastly, it is worth considering the scope for setting up in business with your own horse, particularly in forest operations where difficult terrain makes the use of an animal more practical than machinery. For backing out felled trees and thinnings and general hauling operations, under contract to the Forestry Commission, or private estate engaged in forestry operations, there is already a limited scope. These opportunities will increase in the future and occur also in horticulture. Mr Pinney, is insistent that with our knowledge of genetics we can select or 'design' our horses to fit the needs of the present day, and integrate the horse with tractor operations.

There remains a few more rural work opportunities to describe and so this volume is almost complete! Without exception, the vacancies occuring and the future opportunities are few. What they have in common is their valuable contribution to the rural scene.

The Farrier

Having told you about the work with the heavy horse where better to start the finale of our survey than with the craft of Farriery. The shoeing smith is as essential to the working horse as his fodder, as well as being needed to keep all other horses on the move. The craft is known to every boy and girl through the medium of poetry and it can certainly still be said, 'The smith a mighty man is he, with arms like iron bands.' Should you wish to bend your muslces to this calling, write for: *How to Become a Farrier*—free from *The Worshipful Company of Farriers* 3 Hamilton Road, Cockfosters, Barnet, Hertfordshire EN4 9EU.

Organic Farming

Until the advent of artificial fertilisers, herbicides, fungicides and insecticides, the whole pattern of food-growing was built round the maintenance of the soil structure by returning to it the nutriment taken out by the growing crops. Animal manures and green manure were used, often mixed with straw or composted. What is now achieved by the use of sprays was made possible by hand or horse operations to keep down the competing weeds, while in the process of turning and moving the soil crops could be partially protected from all manner of pests threatening their survival. The natural pattern of one creature living off another tended to help the vegetation prosper.

This book is not the place to debate conventional, modern farming, as compared with the age-old practice or organic farming. What is important is that you should be aware of the conflict of attitudes and find out for yourself what the debate is about. There are limited opportunities to learn organic farming on a practical level and supplement your practical training by college courses devoted to the subject . For full details of what is happening world-wide, as far as organic farming is concerned, and details of job vacancies and for training course details, you should contact: *The Soil Association* Walnut Tree Manor, Haughley Stowmarket, Suffolk, IP14 3RS, who publish a quarterly journal giving details of working opportunities, including those overseas.

Conservation and the environment

Don't be put off by these two words which are very much 'in' nowadays. What they mean is we have to care about the place we live in and be sensitive to the delicate balance of nature. The balance must always be at risk when possibly 55 million people are living in less than 125,000 square kilometres of land mass—that is the size of Great Britain! All of them demanding large amounts of water, high living standards, including the right to own a motor car and to possess adequate accomodation and services.

Food growing, forestry, leisure and recreation all have to be fitted into the overall pattern, so each of us has to consider the strain placed on our total resources so that our children and their children can enjoy what we take for granted as our natural right.

The air we breathe is an invisible but vital part of our environment, as is all the water in our land. Buildings, monuments, bridges and our industrial heritage are also important and you will have read earlier of the role played by the National Trust in conserving many of our historic places. Successive governments have played an important role in trying to balance the conflicting interests to ensure we pursue an adequate conservation policy but in the long run the greatest guarantee we have is aware, alert and active public opinion.

Job opportunites exist as well as a variety of voluntary tasks to be tackled at weekends and during holidays. As yet, there is no career structure if you would like to make conservation and the environment your occupation but there are openings. In the first instance you should write for details of these to: *The Nature Conservancy Council* P.O. Box 6, Godwin House, George Street, Huntingdon, PE18 6BU

Thatching

This century-old occupation calls for a willingness to work outdoors in most weathers, a skill in the use of hand tools and the ability to utilise natural materials, reed, hazel spars, heather and birch spray. Planning laws in some regions of England call for the maintenance of existing thatch used in roofing houses, in fact the thatched house is one of the most pleasing features of our landscape. Should you live in, or near, an area where thatching is carried on and feel inclined to master the craft, write in the first instance to: *The Master Thatchers Association* COSIRA, 35 Camp Road, Wimbledon Common, London, SW19.

However, should you wish to find out more before exploring job opportunites in thatching there is a leaflet (one of a series devoted to *Working in Rural Crafts)* obainable, price ten pence, from: *Careers Occupational Information Centre* The Pennine Centre, 202 Hawley Street, Sheffield, S1 3GA which are the organisation I have already written about in connection with gamekeeping and fish farming.

Continue your quest for jobs on the land. I hope you have found encouragement to join those who make there living from our unique and peerless countryside. When you do so, I wish you every success and can promise you that if you care for the land, and leave it just little a better than you found it, you will obtain job satisfaction beyond the dreams of an urban or town worker.

A forest area: in the foreground, from left to right, Sitka spruce, Scots pine and larch

15 Important fringe jobs

On the fringe of agriculture is a large variety of career opportunities outside the scope of this book but you should be aware of them.

Banking

Banking is now a vital service to farming but the attractions of what we might call Lombard Street, are unlikely to appeal to you if you want a life out of doors.

The farm market

Up and down the land there are weekly markets in most of our county towns. Within the age-old industry they represent auctioneers, valuers and their back-up teams provide much rural employment but, here again, we have come away from the living soil.

The veterinary surgeon

The highly specialised services of the veterinary surgeon too are an essential to all involved in the rearing of animals. The farmer is no exception, as his cows, pigs, sheep, poultry and beef cattle are often dependent on the instant availability of the vet.

The seed merchant

The seed, grain, feedstuff and fertilizer merchant is another national institution although becoming more and more swallowed up in what I will call 'agribusiness'. His is an integral and vital part, with a work force consisting of representatives, office staff and lorry drivers to deliver the seeds and fertilizers and to collect the farmers' produce. An important source of rural job opportunity.

'Agribusiness'

I have mentioned 'agribusiness' and will now briefly explain its nature as it is the greatest single commercial interest involved in farming. The supply of herbicides, pesticides and artificial fertilizers, without which the present high yields of the soil would be impossible, are part of its stock in trade. So, too, is the supply of agricultural equipment and machinery. This can range from the teat cup of a milking machine and its rubber liners to the largest tractor and combine harvester, from a scrubbing brush to a silo the height and girth of a village church or to grain-storage buildings occupying double or treble the floor space of the average village hall.

In this field of activity, the companies involved, mostly internationally based, require a large, qualified staff who must penetrate every corner of Britain to offer their products to the farmer. It is something of a talking point with the farming community to compare the amount of time they have to spend in their long working week meeting with reps in the difficult task of trying to obtain the best item of equipment, as economically as possible for the task to be tackled. With the need to spend up to tens of thousands of pounds on a heavy tractor, combine harvester or new buildings, this can prove to be a daunting task.

The Ministry of Agriculture, Fisheries and Food

This survey of ancillary industries within agriculture would be incomplete without mention of the Ministry of Agriculture, Fisheries and Food. Briefly, the Ministry expresses the will of Parliament as it effects the nation's most important industry, which is given pride of place in its title. From regional bases throughout the country a national advisery service operates. Also, the Ministry sees that the negotiating machinery for wages and conditions runs smoothly.

All employees have the status of civil servants and, where necessary, agricultural experience and qualification is required. Full details of job opportunities can be obtained from the Ministry whose address you will find at the end of this book.

The farm contractor

Mention must be made of the farm contractor who, increasingly, copes with land drainage, fencing, field cultivation, spraying, hay-making, harvesting and milking. Although the use of the contractor is not yet as widespread as in the United States and some European countries, particularly Holland, he will provide an increasing proportion of the labour needs of farming and, as a result, offer career opportunities to the trained and skilled worker at enhanced rates of pay.

The farm secretary

Agriculture, in fact all rural based enterprise, finds itself burdened by more and more paper work. If the farmer or employer were to attempt to cope with this there would be little time left to plan the growing of food or get on with the practical side of work in the country. The Training Board puts this rather more grandly, 'modern business management techniques are being applied more and more to farming today'.

Hence, there are openings for farm secretaries which could involve you in going to a number of small farms and enterprises to help out with paying wages, keeping PAYE records and dealing with the farm accounts and correspondence. You would be employed by an organisation providing farm services. There are also full-time farm secretary posts on the larger farms and in a post of this nature you could be called upon to help out in a variety of ways and certainly would not be in danger of becoming desk-bound!

For this extremely interesting and independent existence you can take a course of training at some of the agricultural colleges listed. This is normally of a year's duration and entry would depend on a good 'O' level examination result.

Having read this book I hope you have a clearer picture of what farming is all about and of the many and varied jobs which contribute to the running of a successful farm. I would emphasise yet again the importance of the Apprenticeship Scheme and of gaining qualifications which will open up to you so many opportunities.

Useful addresses

Agricultural Training Board (Regional Offices)
East Anglia: Phoenix House, 67a High Street, Haverhill, Suffolk, CB9 8AH.
Midlands: 99 Westgate, Grantham, Lincolnshire, NG31 6LE.
Northern: Princes House, 13 Princes Square, Harrogate, North Yorks, HG1 1LW.
South East: 46 London Road, Reading, Berkshire, RG1 5AF.
South West: New Oxford House, 9 East Street, Taunton, Somerset, TA1 3LL.
Wales and Border Counties: 45 High Street, Shrewsbury SY1 1ST.
Scotland: 13 Marshall Place, Perth PH2 8AH.

Agricultural Wages Board, Eagle House, Cannon Street, London EC4.

Association of Agriculture, Victoria Chambers, 16/20 Strutton Ground, London SW1P 2HP.
(encourages the teaching of young people about the importance of agriculture and the land).

Bee-keeping:
British Bee-keepers Association, 55 Chipstead Lane, Riverhead, Sevenoaks, Kent.
Central Association of Bee-keepers, Longreach, Stockbury Valley, Sittingbourne, Kent.

Conservation and the Environment:
The National Trust, 42 Queen's Gate, London SW1.
The Conservation Society, 12a Guildford Street, Chertsey, Surrey, KT16 9BQ.
Country Landowner's Association, 16 Belgrave Square, London SW1X 8PQ.

Contract work: The National Association of Agricultural Contractors, Huts Corner, Tilford Road, Hindhead, Surrey, GU26 6SF.

Colleges of Agriculture and Horticulture

Bedfordshire: Shuttleworth Agricultural College, Old Warden Park, Biggleswade.

Bedford College of Higher Education, College Farm, Silsoe.

Berkshire: Berkshire College of Agriculture, Hall Place, Burchett's Green, Maidenhead SL6 6OR.

Buckinghamshire: Aylesbury College of Further Education and Agriculture Department of Agriculture and Horticulture, Hampden Hall, Stoke Mandeville HP22 5TB.

Cambridgeshire: Isle of Ely College of Further Education and Horticulture, Ramnoth Road, Wisbech PE13 2JE.

Cheshire: Cheshire College of Agriculture, Reaseheath, Nantwich CW5 6DF.

Cumbria: Cumbria College of Agriculture and Forestry, Newton Rigg, Penrith.

Derbyshire: Derbyshire College of Agriculture, Broomfield, Morley, Derby DE7 6DN.

Devon: Seale-Hayne College, Newton Abbot.

Bicton College of Agriculture, East Budleigh, Budleigh Salterton, Devon, EX9 7BY.

Dorset: Dorset College of Agriculture, Kingston Maurward, Dorchester DT2 8PY.

Durham: Durham Agricultural College, Houghall, Durham, DH1 3SG.

East Sussex: Plumpton Agricultural College, Plumpton, Lewes BN7 3AG.

Essex: Writtle Agriculture College, Writtle, Chelmsford CM1 3RR.

Gloucestershire: Royal Agriculture College, Cirencester.

Gloucestershire College of Agriculture, Hartpury House, Gloucester GL19 38D.

Hampshire: Hampshire College of Agriculture, Sparsholt, Winchester.

Hereford and Worcester: Pershore College of Agriculture, Avonbank, Pershore WR10 3JP.

Hertfordshire: College of Agriculture and Horticulture, Oaklands, St. Albans.

Humberside: Bishop Burton College of Agriculture, York Road, Bishop Burton, Beverley HU17 89G.

Kent: Hadlow College of Agriculture and Horticulture, Hadlow, Tonbridge.

Lancashire: College of Agriculture, Myerscough Hall, Bisborrow, Preston PR3 0RY.

Colleges of Agriculture and Horticulture *(continued)*

Leicestershire: Brooksby Agricultural College, Brooksby, Melton Mowbray LE14 2LJ.

Lincolnshire: Kesteven Agricultural College, Caythorpe Court, Grantham NG32 3EP.

Lindsey College of Agriculture, Riseholme, Lincoln LN2 2LG.

Norfolk: Norfolk College of Agriculture and Horticulture, Easton, Norwich NR9 5OX.

Northamptonshire: Northamptonshire College of Agriculture, Moulton, Northampton NN3 1RR.

Northumberland: Northumberland College of Agriculture, Ponteland, Newcastle upon Tyne NE20 0AQ.

North Yorkshire: Askham Bryan College of Agriculture and Horticulture, Askham Bryan, York YO2 3PR.

Nottinghamshire: Nottinghamshire College of Agriculture, Brackenhurst, Southwell NG25 0QF.

Oxfordshire: Rycotewood College, Priest End, Thame OX9 2BR.

North Oxfordshire Technical College and School of Art, Broughton Road, Banbury.

West Oxfordshire Technical College, Holloway Road, Witney, Oxon 0XB 7EE.

Salop: Harper Adams Agricultural College, Newport TF10 8NB.

Shropshire Farm Institute, Walford, Baschurch, Shrewsbury SY4 2HL.

Somerset: Somerset College of Agriculture and Horticulture, Cannington, Nr. Bridgwater TA5 2LS.

Staffordshire: Staffordshire College of Agriculture, Rodbaston, Penkridge, Stafford ST19 5PG.

Suffolk: Chadacre Agriculture Institute, Chadacre, Shimpling, Bury St. Edmunds, Suffolk, IP29 4DU.

East Suffolk College of Agriculture and Horticulture, Otley, Ipswich, Suffolk, IP6 9EY.

Surrey: Merrist Wood Agriculture College, Worplesdon, Nr. Guildford GU3 3PE.

Warwickshire: Warwickshire College of Agriculture, Moreton Hall, Moreton Morrell, Warwick CV34 9BL.

Wiltshire: Lackham College of Agriculture, Lacock, Chippenham SN15 2NY.

Wales:
Clwyd: Llysfasi College of Agriculture, Ruthin, Clwyd LL15 2LB.
The Welsh College of Horticulture, Northop, Nr Mold, Clwyd CH7 6AA.
Dyfed: Carmarthen Technical and Agricultural College, Pibwrlwyd, Carmathen.
Gwynedd: Glynllifon College of Further Education, Clynnog Road, Caernarvon LL54 5DU.
Gwent: The Usk College of Agriculture, Usk NP5 1XJ.
Mid-Glamorgan: Mid-Glamorgan College of Agriculture and Horticulture, Pencoed, Bridgend.
Powys: Montgomery College of Further Education, Newton.
Joint Education Committee: Welsh Agricultural College, Llanbadarn Fawr, Aberystwyth, Dyfed.

Complete catalogue of Books on Agriculture, Horticulture, Forestry and Allied Subjects available from: Landsman's Bookshop Ltd, Buckenhill, Bromyard, Herefordshire.

Commercial and Amenity Horticulture: Careers in park and recreation administration: Write: The Secretary, Institute of Park and Recreation Administration, Lower Basildon, Reading, Berks, RG8 9NE.

Council for Nature, Zoological Gardens, Regents Park, London NW1.

Engineering (Agricultural) For career details and apprenticeship scheme write: The Education Officer, National Joint Apprenticeship Council for the Agricultural Machinery Trade, Church Street, Rickmansworth, Herts.
or: The Institution of Agricultural Engineers, West End Road, Silsoe, Bedford MK45 4DU.

International Work Opportunities: write: *Voluntary Service Overseas,* 9 Belgrave Square, London SW1X 8PW.
International Voluntary Service, Section OS52, 53 Regent Road, Leicester LE1 6YL.

Land Settlement Association 43 Cromwell Road, London SW7.

Manpower Services Commission Selkirk House, 166 High Holborn, London WC1V 6PF.
(for details of Youth Opportunities Programme in agriculture for the young unemployed).

National Farmers Union: write to Information Officers
Cumbria, North Riding and Durham, Northumberland, Yorkshire (E. Riding), Yorkshire (N. Riding): Agriculture House, Salters Lane South, Haughton-le-Skerne, Darlington, Co Durham DL1 2AA.

Cheshire, Derbyshire, Lancashire, Yorkshire (West Riding): Agriculture House, 119 Nantwich Road, Crewe, Cheshire CW2 6BD.

Herefordshire, Leicestershire, Northamptonshire, Rutland, Nottinghamshire, Oxfordshire and Berkshire, Shropshire, Staffordshire, Warwickshire and Worcestershire: 49-57 High Street, Droitwich, Worcs.

Bedfordshire and Huntingdonshire, Buckinghamshire, Cambridgeshire, Essex, Hertfordshire, Lincolnshire, Norfolk, Suffolk: Agriculture House, George Street, Huntingdon.

Cornwall, Devon, Dorset, Gloucestershire, Somerset, Wiltshire: Agriculture House, 31 Trull Road, Taunton, Somerset TA1 4QG.

Central Southern, Hampshire and Isle of Wight, Kent, Sussex: Agriculture House, Station Road, Liss, Hampshire GU33 7AR.

Wales and Monmouth, Anglesey, Brecon and Radnor, Cardiganshire, Caermarthenshire, Denbigshire, Flintshire, Glamorgan, Merioneth, Mid Gwynedd, Monmouthshire, Montgomeryshire, Pembrokeshire: Agriculture House, 19/21 Cathedral Road, Cardiff CF1 9LJ.

National Proficiency Tests Council YFC Centre, National Agricultural Centre, Kenilworth, Warwickshire CV8 2LG.

National Union of Agricultural and Allied Workers Headland House, 308 Gray's Inn Road, London WC1X 8D.

Useful addresses

Royal Agricultural Society of England National Agricultural Centre, Kenilworth, Warwickshire CV8 2LZ *also supply career information.*

The Soil Association Walnut Tree Manor, Haughley, Stowmarket, Suffolk IP14 3RS.

Young Farmers' Clubs, YFC Centre, National Agricultural Centre, Kenilworth, CV8 2LG and finally some Government addresses:
Department of Education and Science (Information Division), Elizabeth House, York Road, London SE1 7PH.
For details of Universities providing degree courses

Department of Employment Training Services Agency, Ebury House, Ebury Bridge Road, London SW1.

Ministry of Agriculture, Fisheries and Food, Whitehall Place (West Block), London SW1A 2HH.

Ministry of Agriculture, Fisheries and Food, Establishment Division, 3-34 Victory House, Kingsway, London WC2B 6TU.

Department of Agriculture and Fisheries for Scotland, Chesser House, Gorgie Road, Edinburgh EH11 3AW.

Note The local office of the Ministry of Agriculture and of the Forestry Commission can be found in your local telephone directory.

Index

Index